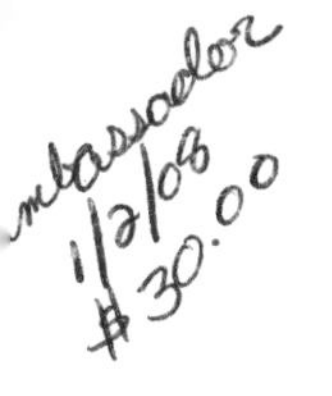

W9-CDO-885

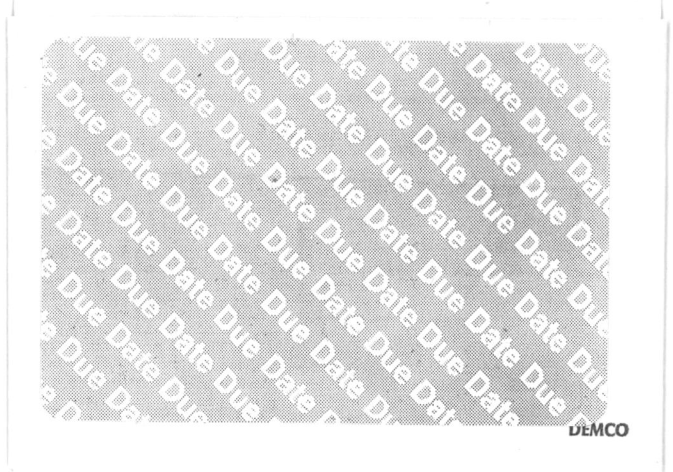
DEMCO

Breathing Space

Breathing Space

How Allergies Shape Our Lives and Landscapes

GREGG MITMAN

YALE UNIVERSITY PRESS / NEW HAVEN & LONDON

Portions of material in chapters 1, 2, and 3 of this book were published in different form in the following:

"Hay Fever Holiday: Health, Leisure, and Place in Gilded-Age America," *Bulletin of the History of Medicine* 77 (2003): 600–635.

"Natural History and the Clinic: The Regional Ecology of Allergy in America," *Studies in History and Philosophy of Biological and Biomedical Sciences* 34 (2003): 491–510.

"When Pollen Became Poison: A Cultural Geography of Ragweed in America," in *The Moral Authority of Nature*, ed. Lorraine Daston and Fernando Vidal (Chicago: University of Chicago Press, 2004), 438–465.

"Geographies of Hope: Mining the Frontiers of Health in Denver and Beyond, 1870–1965," *Osiris*, 2nd ser. 19 (2004): 93–111.

Designed by Mary Valencia.
Set in Stone by Keystone Typesetting, Inc. Printed in the United States of America.

Library of Congress Catalog Card No. 2006052827
ISBN: 978-0-300-11035-7 (cloth : alk. paper)

A catalogue record for this book is available from the British Library.

The paper in this book meets the guidelines for permanence and durability of the Committee on Production Guidelines for Book Longevity of the Council on Library Resources.

10 9 8 7 6 5 4 3 2 1

To my parents

CONTENTS

PREFACE

Crystalline snowflakes blanket Pennsylvania's Lehigh Valley in a fresh layer of late-winter snow and spread a hush over my fifth hospital season. I am five years old, and death had come too close this time. In my hospital bed I am enveloped in a translucent plastic tent that muffles visitors' voices and amplifies my breathing, seemingly in concert with the steady rhythm of the machine pumping water vapor and oxygen into the cool, misty space around me. My mother hovers, relieved that my once blue lips—a sign of how poorly oxygenated my blood had become—are pink again. My father, holding back tears, steps out into the hall to smoke a cigarette and ease his nerves.

This is my most vivid childhood memory—the quiet encapsulation of what was thankfully the last time I needed the engineered atmosphere of a hospital oxygen tent to breathe. As a child with bronchial asthma growing up in the white, middle-class suburbs of 1960s America, I took for granted oxygen tents, allergy clinics, and dust-free bedrooms with hardwood floors. Allergy became my way of being in the world. I learned to avoid certain places—the home of a favorite aunt with a pet parakeet, a nearby farm where my sister rode horses—while other places, such as a twin bed in the cool, damp basement of our hilltop home, offered refuge. Some spaces were remade as the doctors advised: a vaporizer became a fixture of my bedroom, while stuffed animals were banned.

In that last hospital season of 1966, I was one of an estimated 12.6 million people with allergies in the United States. Today, allergic asthma and hay fever affect more than 50 million Americans.[1] With overly sensitive immune systems that react too strongly to the world around us, we allergy sufferers have special powers of perception. Depending on one's particular sensitivities, some of the tiniest natural and man-made substances in the environment can be noticed. Pollen grains, less than thirty microns in size, are innocuous to most people, but in sensitive sufferers they can let loose a torrent of runny noses, watery eyes, sneezing fits, and asthma attacks. The slightest trace of peanuts will send an allergic child into anaphylactic shock, a life-threatening allergic overreaction. A bee sting, the touch of latex, or a dose of penicillin can quickly kill the too sensitive. Smog hanging over the valley of an industrial town, the toxic effects of ozone, and any number of particulates in the air can inflame the respiratory passages of asthmatics, many of whom rely on still more chemicals, delivered to the lungs through the puff of an inhaler, to open up their airways and aid in their struggles to breathe.

To the roughly 20 percent of Americans who are allergy sufferers, the details of environment and place figure centrally in our lives. We learn to take special precautions as we make our way through the world: an albuterol inhaler or epi pen kept always handy; vacation getaways timed to the release of a dreaded pollen at home; strict attention to the labeling on packaged food—"May contain peanuts"—that can save a life. This has been also true for those afflicted by allergies ever since the mid-nineteenth century. The story of my childhood, like that of countless other sneezing and wheezing individuals of the past, is part of a larger history of place and transformation.

Allergy, from its appearance in the 1870s, first among the wealthier classes, has been understood as a disease of progress, a symptom and a sign of the modern conditions in which Americans live. Hay fever sufferers of the Gilded Age regarded their affliction as an indicator of the nervous exhaustion that accompanied modern civilized life. Heightened sensitivity to climatic change and a host of external irritants, such as dust, sunlight, and plant pollens, they reasoned, was evidence of a refined sensibility cultivated by superior intellect, culture, and wealth. After World War II, physicians such as Warren Vaughan, who promoted allergy as a "malady of civilization," described modern industrial and technological development, and in particular the flood of new chemicals then being introduced into the environment, as contributive causes.[2] Within the last two decades, it is not impure civilization but the too hygienic conditions of modern life that allergists and journalists have seized upon as an explanation for the increased prevalence of immune system hypersensitivity and allergy. Yet Americans in the twentieth and twenty-first centuries have been largely unaware of links between the changing conditions of modern life and the growth of allergic disease. The population shift from rural to urban, the building of railroad networks, movement of plants and people westward, growth of industrial cities, and increased time spent indoors have each played a role. Moreover, the place of allergy in the making of the American landscape has remained largely hidden from view. As different allergens have come into being—the stresses of civilization, pollens, cockroaches, air pollutants, molds, and dust mites, for instance—we have modified the spaces where we live, work, and play, hoping to breathe more easily. But often we have exacerbated the allergic landscape and made worse the very symptoms that we have aimed to relieve.

In the last fifty years, the emphasis on drugs in the treatment of allergy has focused attention on the body isolated from its environment. Antihistamines, bronchodilators, and corticosteroids, introduced after World War II, offered Americans effective relief from their allergy symptoms and made physicians optimistic about their ability to engineer the body's immune terrain and control allergy's rise.

As an asthmatic child, I benefited greatly from new drug therapies introduced in the postwar era. They probably saved my life. Still, while physicians were eager to prescribe Chlor-Trimeton, prednisone, or whatever latest drug appeared on the market, the emphasis on drugs skewed the attention paid to the places and spaces in which we live. My doctors asked, for example, more about the emotional environment of our home—a patient's emotional state was a widely suspected cause of asthma in the early 1960s—than about the smoking habits of my father and several of my uncles.

My good fortune, that of growing up in a family able to make the necessary sacrifices to get quality health care, made my experience of illness and environment very different from that of my brother-in-law, Akheem Torres, who grew up in New York City, only a two-hour drive away. In 1956, as a four-year-old child, Akheem came to the United States on a Pan American flight from San Juan to New York City. He was one among thousands of Puerto Ricans who immigrated to the United States in the 1950s, attracted by the promise of better jobs and a better life. His family lived in a spacious apartment in Brooklyn, where Pops, his stepfather, served as a superintendent in exchange for free rent. When Pops died of stomach cancer six years later, at age thirty-six—a death Akheem now attributes to contaminants in the transistor radio factory where his stepfather worked—the family became trapped in a downward economic spiral. At thirteen, Akheem moved with his mother and two sisters into a two-room apartment in an industrial neighborhood on Manhattan's Lower East Side that was infested with some of the "biggest-ass rats" and roaches he had ever seen. Akheem escaped to the rooftop, where he inhaled the sulfurous haze (from apartment incinerators and coal-fired plants) that hung over building tops and added brilliant reds to the sunsets he watched from the tar beaches. "You couldn't keep people in bricks and mortar," he told me. Everybody needed to claim some little piece of the outdoors, some little space where it was possible to breathe. Akheem's younger sister, Millie, had a more difficult time finding a space to breathe. Respiratory infections, allergies, and asthma plagued her childhood. Akheem can still vividly recall how they would fabricate a homemade vaporizer when Millie gasped for breath: they would drape a blanket over both her head and a pot of simmering water laced with Vicks VapoRub, the smell of which wafted through the air. Only in a dire emergency would Akheem's family have thought of taking her to a doctor or hospital. Without health insurance and with little income, the family had to make do on its own.

In the different spaces in which Akheem and I lived and breathed, asthma became two quite different diseases. Because asthma and hay fever have many different ecologies, these diseases are experienced in different places and times in different ways. These diseases are the result of complex interactions with physical, chemical, biological, and social factors in the environment. We can continue to grope at individual causes and remedies, one after another, hoping to fortify our bodies against an ever-changing environment with an ever-changing barrage of drugs, while neglecting the environmental and health disparities that accompany economic progress. Or we can learn from history and do better. We can widen our focus on the causes and prevention of allergic illness and invest more in research that takes into account the ecological relationships between illness and place that have structured the experience and rise of allergy in America. The choice is ours.

Numerous people and organizations helped create an intellectual environment and generously offered valuable time and resources to make this book a reality. Initial support for this project came from NSF grant SES-0196204. A year at the Max Planck Institute for the History of Science in Berlin, with a fellowship from the Alexander von Humboldt Foundation, put me in contact with a wonderful cadre of fellow scholars and friends who helped shape this project in its formative stages. My thanks to Lorraine Daston for making it all possible. An international conference on Environment, Health, and Place in Global Perspective, which I co-organized with Michelle Murphy and Christopher Sellers and which was funded by NSF grant SES-0114570 and the University of Wisconsin's Anonymous Fund, Gaylord Nelson Institute for Environmental Studies, Department of Medical History and Bioethics, Department of History of Science, and Robert F. and Jean E. Holtz Center for Science and Technology Studies, raised many challenging questions at the intersections of environmental and medical history; I have sought to address such questions in this book. I am grateful for a year spent at the National Humanities Center, with additional support from the American Council of Learned Societies and a fellowship from the John S. Guggenheim Memorial Foundation; it came at a fortuitous time in the final stages of writing. At the National Humanities Center, my thanks to fellow historians Julia Clancy-Smith, Ed Curtis, Deborah Harkness, Lisa Lindsay, Cara Robertson, and Tim Tyson for taking the time to read and share comments on each other's work.

Closer to home, I feel particularly fortunate to have a group of supportive colleagues and friends at the University of Wisconsin, in both the Department of Medical History and Bioethics and the Department of History of Science. The Gaylord Nelson Institute for Environmental Studies has been an especially welcoming and fertile place for interdisciplinary research. Many thanks to all the faculty and graduate students who have made the Culture, History, and Environment Research Circle a friendly and stimulating meeting ground to discuss and debate the many ways to think about and write environmental history.

Without the research assistance of Victoria Elenes, Paul Erickson, Maureen McCormick, Erika Milam, and Camilo Quintero this book would have taken much longer. Thanks for all your hard work and creative ideas. The library staffs at the following institutions gave much of their time and knowledge: the Arizona Historical Society; the Archives of the American Academy of Allergy, Asthma, and Immunology at the University of Wisconsin, Milwaukee; the Bentley Historical Library at the University of Michigan; the Bancroft Library at the University of California–Berkeley; the Bay View Archives at Bay View Association, Michigan; the Rare Books and Special Collections Department at the Countway Medical Library of Harvard University; the Denver Public Library;

Special Collections and Archives at the Penrose Library of the University of Denver; Mackinac State Historic Parks; the History of Medicine Division of the National Library of Medicine; the New Hampshire Historical Society; and the University of Arizona Historical Collections. Alice Holst kindly provided me with the correspondence and unpublished manuscripts of Oren Durham. Micaela Sullivan Fowler, Director of Historical Services of Ebling Library at the University of Wisconsin–Madison, has always been enthusiastic and gracious whenever I've barged into her office with a research request. Thanks to her for always making a visit to Ebling Library such a delight.

Other scholars have generously shared their knowledge about medical history, the history of allergies, and disease ecology. First and foremost, my thanks to Sheldon Cohen for his gracious hospitality every time I've visited Washington, D.C., and for his remarkable insights into the history of allergy as a clinical field. Conversations with Warwick Anderson, Sara Hotchkiss, Mark Jackson, Rick Keller, Harry Marks, Linda Nash, Jonathan Patz, and Christopher Sellers have always proven invaluable at various stages along the way. Thanks to Ken Adler and John Tresch for their knowledge of Lake Forest and Mr. T.

I also feel incredibly grateful to have assembled such an expert team of readers and editors. Ronald Numbers and Michelle Murphy read every page of the manuscript and offered invaluable comments. Few scholars are blessed with such fine critics and wonderful friends. Lisa Adams, my agent, has invested more time in me and in this project than I ever expected. Sarah Flynn jumped in at a critical time and improved the manuscript in countless ways. Bojana Ristich was the best copyeditor any writer could hope for. Carol Dizack worked wonders with the illustrations for the book. My editor at Yale University Press, Jean Thomson Black, has been an enthusiastic coach and helpful guide. Susan Lampert Smith has a keen eye for a good story; how fortunate to have a journalist and fellow asthmatic friend in my midst. Thanks also to Peter Townshend, who generously granted permission to reprint lyrics from The Who's 1971 classic "Going Mobile."

More than any of my previous books, this one touches family. Until I wrote this book, Mom, I hadn't quite realized how much you and Dad endured and gave to make a healthy space for your son. How can I ever repay the sacrifices you, Dad, Denice, and Robyn made? To Akheem, you are my brother now, no matter how distant the spaces of our past lives.

To Debra, you have given more of your time, knowledge, love, and fine editorial hand than any other person. Each day I awake next to you is a breath of fresh air. To Keefe, what a joy you bring. I only hope my generation can bequeath to your generation a more ecologically just planet on which to breathe.

Breathing Space

Introduction

The scream of chain saws gnawing through full-grown oak trees abruptly announced the start of summer in 1987 in the leafy enclave of Lake Forest on the Lake Michigan shore north of Chicago. Most residents treasured the trees that shaded their mansions, but Mr. T wasn't like most of his neighbors. Laurence Tureaud, known to most Americans as Mr. T, star of the 1980s hit TV show *The A-Team* and Rocky's nemesis in *Rocky III,* had moved into the exclusive, historic neighborhood the previous fall, when the hundreds of oaks and other native plantings at Two Gables, his seven-acre estate, lay dormant. That spring, as the trees on the property came into bloom, the six-foot-tall, 220-pound, well-muscled actor, who had survived growing up in the grinding poverty of Chicago's largest ghetto, suffered terribly. Pity the fool, or in this case the tree, that tangled with Mr. T. Saying that allergies to tree pollen provoked him, the man who wore golden chains as a reminder of the chains of his slave ancestors, took action. With a chain saw in hand and the help of hired workers, Mr. T cleared the property of more than seventy trees, many of which dated back to the early twentieth century and had been planted under the direction of the famed landscape architect Jens Jensen.

The "Lake Forest Chain Saw Massacre" didn't win Mr. T any favor in his new neighborhood, which had won the label of Tree City, USA, for seven consecutive years from the National Arbor Day Foundation.[1] Determined that no one should try again to alleviate any ill with a chain saw, the Lake Forest city council passed an ordinance making it illegal for a private homeowner to remove large trees within thirty-five feet of public thoroughfares without a permit (Figure 1).[2]

Figure 1. When Mr. T took a chain saw to remedy his reported allergies to tree pollen, Lake Forest residents were up in arms. Courtesy of *Chicago Tribune.*

Angry neighbor Betsy Kitzerow told *Jet* magazine she couldn't understand why Mr. T "didn't buy himself a high rise and surround himself with concrete" rather than be "destructive to something that took God maybe 60 years to grow."[3] But Lake Forest wasn't made by God. It was a landscape shaped in the late nineteenth and early twentieth centuries by wealthy industrialists, financiers, and businessmen who were, like Mr. T, concerned with health.

What Mr. T's neighbors didn't appreciate, or care to recognize, is that the places where Americans have struggled to breathe, as well as the spaces they have created to breathe more freely—lavishly landscaped estates, hay fever resorts, air-conditioned homes—have been shaped, not only by the ecology of animal, insect, plant, and man-made allergens, but also by the unequal distribution of wealth and health care in American society. And the environmental and economic disparities associated with allergy in America are no less today than they were in the nineteenth century, when allergy first made a visible mark upon the North American landscape. Just as wealth afforded nineteenth-century Americans the luxury of escaping the environmental conditions of modern industrial life, money today (translated as health care benefits and the relative freedom to choose where to live) gives Americans who have it the ability to reduce their exposure to the environmental burdens of modern society.

When Mr. T moved into Two Gables, he had finally turned the tables. His historic ten-thousand-square-foot mansion, with a driveway a block and a half long, was far upwind of the industrial pollutants, pesticides, poverty, and violence that he had dreamed of escaping as a youth. Now, he said, he was the "black version of the Beverly Hillbillies." Born in Cook County Hospital in 1952, Laurence Tureaud had grown up on Chicago's South Side. His mother, who single-handedly raised twelve children, moved the family into the Robert Taylor Homes when Laurence was ten.[4] Opened in 1962, the twenty-eight separate buildings (each sixteen stories) stretched over two miles in the shadow of the Dan Ryan Expressway and represented the single largest public housing project in the world. Under the legendary mismanagement of the Chicago Housing Authority, the Robert Taylor projects also became home to the largest concentration of poverty in America. Degraded housing stock, high levels of environmental pollutants, and the residents' inadequate access to health care and lack of financial resources turned this space and other public housing projects across urban America into asthma zones. By the 1980s, severe asthma had increased disproportionately among poor children of color across America's urban landscape. America's urban minority and poor populations had increasingly borne the environmental and social costs of modern society.

In the public housing projects of Chicago, African Americans were almost five times more likely to die of asthma than whites.[5] As a child, Mr. T had little control over his environment. As an adult, armed with a chain saw and empowered with money he had earned as a Hollywood star, he, like wealthy hay feverites of the nineteenth century, had the ability and luxury to alter his environment as he saw fit. Though some of his neighbors thought he didn't fit in—and Mr. T might have agreed—he was, in fact, more akin to the founders of Lake Forest than any of them realized. In the 1850s, a group of Chicago Presbyterians wanted to establish a residential community, united through education and worship, that would also offer an escape from the dangers and diseases that modern civilization seemed to breed. They founded Lake Forest thirty miles north of Chicago, atop the bluffs along Lake Michigan's western shore, where dense stands of oak and patches of open prairie caught the cool, refreshing lake breezes.[6]

By 1869, with the opening of the transcontinental railroad, Chicago had become a thriving commercial metropolis, the center of an empire of bountiful natural resources—iron ore, timber, livestock, and grain—transported raw or as finished products and traded as commodities on the Chicago Exchange.[7] Wealth accumulated in the hands of entrepreneurs, including Gustavus Swift, Philip D. Armour, and Cyrus McCormick, who transformed American agriculture into an industrial enterprise. The rapid movement by railroad of capital, goods, and people across widespread regions that had been relatively isolated resulted in new anxieties and illnesses. Outbreaks of cholera and epidemics of typhoid fever, the influx of immigrants, and the sight and stench of garbage on city streets all propelled the city's upstanding Protestant business leaders to seek refuge for their families away from the emerging threats of the urban environment.

Hay fever was one such new threat. Known to its nineteenth-century sufferers as June cold, rose cold, hay asthma, hay cold, or autumnal catarrh, hay fever first appeared in America as an upper-class disease around the time Lake Forest was founded. Affecting the wealthy, educated, urban elite who had the means and need to establish communities of refuge, hay fever was widely regarded as "the price of wealth and culture, a part of the penalty of fine organization and indoor life."[8] American hay fever sufferers in the Gilded Age, known as hay feverites, saw the rapid progress, moral complacency, and physical degeneracy of modern civilized life as instrumental in bringing about their dreaded disease. For sufferers among America's intellectual, business, and political leaders, hay fever signaled nervous exhaustion. They blamed their "tired, nerve-shaken" bodies on the "poisonous struggling and excitement of city life" and believed their lives were out of place in the modern urban environment that they nevertheless had helped to create and upon which they depended.[9]

Where might relief from hay fever—this malady of civilization—be found? Unlike Mr. T, America's nineteenth-century allergy sufferers looked to nature not as the problem but as the cure. Nature, they believed, provided an antidote to the "special enervating conditions" of the city.[10] On their suburban estates or in fashionable mountain and lakeshore resorts, the affluent afflicted in the nineteenth century shaped the environment in ways they thought would alleviate their symptoms. Aiding them in this endeavor were some of America's leading landscape architects—Frederick Law Olmsted, Charles Eliot, Jens Jensen, and Warren Manning—who, in their designs of public parks and country gardens, hoped to create "breathing places of various sorts" that would offer relief from the "confined and vitiated air of the commercial quarter."[11]

In 1910, Orville Elias Babcock, a Chicago investment banker, hired pioneer landscape designer Jens Jensen to create a bucolic setting for a traditional English manor home in Lake Forest that he called Two Gables. Like Frederick Law Olmsted (probably best known for designing New York's Central Park) and others in the "natural style" tradition of American landscape architecture, Jensen believed that natural surroundings provided healthful benefits. He rejected the "social display and ostentatious wealth" of fashionable formalism and championed what became known as the prairie style of landscape architecture. Through the use of native plants and the "careful manipulation of colors, textures, sunlight, and space," Jensen sought to capture the spirit of the natural landscape, immersing urban residents in the "cycles of nature around them."[12]

Of course, the natural spaces created so attentively a century before also let loose the pollen that so bedeviled Mr. T and other allergy sufferers. Allergy now fuels a $5 billion industry in antihistamines alone—not to mention other drugs, home care products, and foods and cosmetics targeted at allergy sufferers. Allergy's rise in places like Lake Forest and across America is directly connected to changes in the natural and built environment over the last 150 years. From the development of places like Lake Forest or mountain resort communities in the White Mountains and the Adirondacks to the establishment of national forest reserves and encompassing the settlement of western towns such as Tucson and Denver, allergy has significantly shaped economic development, attitudes toward land use, and the daily lives of Americans.

Environmental historians look upon a landscape and see events and processes—both human and natural—that have molded the terrain over time. For example, the casual spectator touring southwestern Wisconsin may see little more than rolling hills and pasture with a smattering of dairy farms. But to the aware observer, the undulating ridges and valleys and black, sandy loam soils tell a story with many chapters: glacial forces; fires set by lightning strikes or Native Americans; bison grazing;

patterns of human settlement and migration. Disease, too, is a force of human and natural history that leaves its marks on the landscape. A diver in the Mackinac Island harbor of Lake Michigan finds a hay fever elixir bottle dating back to the late nineteenth century. Its discovery reveals a piece of the story of this Great Lakes tourist destination as a nineteenth-century hay fever refuge. Why do asthma and hay fever rates in Tucson far exceed the national average? Look at the Southwest's history as a last resort for allergy and asthma sufferers. When and why did pollen counts become a part of daily weather reporting? Examine the establishment of allergy clinics in the United States and the links between physicians and botanists in the development of early treatments.

Escape was the first remedy Americans sought in their battle against allergic disease. Beginning in the 1870s thousands of wealthy hay fever sufferers fled the heat, filth, and dust of the cities in mid-August each year to lounge at luxurious hay fever resorts in the cultivated wilderness of the White Mountains in New Hampshire, the Adirondacks in upstate New York, and the Great Lakes shores of the Midwest. But as these places gained popularity for being allergy free, the very efforts to escape, if not cure, allergy were accompanied by unintended environmental consequences. The economic development of the White Mountains region as a hay fever resort, for example, depended on the building of railroad lines, which made travel to the region fast and comfortable. Soon thereafter, ragweed, an allergen previously unknown to the region, began to flourish in the disturbed soils along the new railroad corridors. In the 1950s, when thousands began streaming to Tucson in search of relief from hay fever and asthma, the very causes of those allergic diseases traveled west with them. Non-native plants, such as Bermuda grass, began to be seen in Tucson for the first time, and desert weeds proliferated, as did automobiles and industry. Even as health seekers embraced Tucson as a promised land, they longed for the trees and grasses of more northern landscapes. The imported olive and mulberry trees and lawn grasses planted by Tucson homeowners had increased the area's pollen load tenfold by the 1970s. By then, Tucson had a much higher incidence of asthma and allergic rhinitis (hay fever) than the national average, a consequence of both the number of allergy sufferers who had moved there seeking relief and the changes to the environment they had wrought. Here, as elsewhere in America, allergy sufferers hoping that a change of place would be a simple solution to their disease found that civilization, the perceived cause of their ailment, always followed close behind.

As a result, Americans increasingly turned to altering the environments of the city and home to fight the disease. Weed ordinances and eradication efforts were the focus of major public health campaigns in metropolitan centers east of the Rockies by the 1930s. During the Great Depression, New York City put the unemployed to work through the Works Progress Administration to rid the urban landscape of ragweed. They mowed and cut 132,600,000 square feet in the city alone. A noxious enemy of the allergy prone, ragweed is a late summer plant that in the 1930s spewed 275,000 tons of toxic dust into the atmosphere each year. In the 1940s, sanitation engineers gained a seemingly ideal new weapon in their war on allergy-provoking weed: the herbicide 2, 4-D, an invention of the nation's military-industrial complex. By 1955, in the state of New Jersey alone more than 150 communities had adopted municipal ragweed spraying campaigns that used 2, 4-D in efforts to control hay fever. Such programs helped push the sale of the herbicide to more than $50 million that year.

The increased technological optimism that made Americans confident in their ability to rid the landscape of allergy also spurred the growth of the air-conditioning industry. Beginning in the 1930s, air-conditioned homes and buildings were promoted and sold by manufacturers and environmental engineers, who sought to recreate an "artificial climate as nearly as possible like the refreshingly cool and pollen-free atmosphere of the Rocky Mountain and the Northern Great Lake States."[13] But the move to the great indoors carried its own health risks, albeit unanticipated, and these became increasingly visible in the early 1980s. With urban residents spending about 90 percent of their time in indoor environments and with construction trending toward more airtight, energy-efficient buildings in the wake of the energy crisis, indoor contaminants emerged as a national health concern. Such contaminants included passive tobacco smoke, radon, carbon monoxide, nitrogen dioxide, formaldehyde, dust mites, and other aeroallergens. This increase in indoor air pollution fostered yet another lucractive industry—home care and ventilation products aimed at fixing the unintended environmental consequences that resulted from the very efforts to manufacture an indoor breathing space for allergy sufferers.

Accompanying these efforts to alter the external environment was the hope that medicine could change the body's interior environment to insulate allergy sufferers from the increasing presence of recognized outdoor and indoor allergens. And in the 1950s, advances in biomedicine made it possible to believe that drugs and chemicals could provide permanent relief. But environmental illnesses have not yielded readily to the sort of biomedical model that triumphed over killers such as cholera and consumption in the nineteenth century. By looking within, we have lost sight of the environmental and social factors that have shaped our century-long battle with allergic diseases such as asthma and hay fever.

The heralded success of the germ theory of disease in the late nineteenth century, when heroic figures like Robert Koch and Louis Pasteur firmly established the link between infectious diseases and microbes, helped shift the focus from the environment to the individual as the carrier of disease. Bodies became universalized and standardized; drugs and vaccines worked their miracles regardless of the economic status, gender, or race of individuals or the environments in which they lived. Allergy, however, is not easily universalized because it is so much a disease of place. Allergenic pollens and molds, for example, thrive only in some places; air contaminants are different in number, amount, and kind in one area than in another. This biogeographical link means that the problems a poor, asthmatic child in the inner city faces are far different from those of a middle-class, asthmatic child in the suburbs. In our quest for a magic bullet to cure allergic illness, we have sought simple causes and easy solutions. But while biomedicine has ignored the role of the environment beyond the body, allergic diseases have stubbornly and increasingly persisted, yielding an inestimable cost in drug side effects, health care dollars, and both social and economic consequences.

In the case of asthma, which costs the U.S. economy at least $14.5 billion per year, most federal research dollars in 1999 were spent on biomedical research aimed at treatment. Less than 10 percent of spending was targeted at tracking and monitoring programs that could better reveal when and where asthma occurs and its links to potential environmental exposures. In 2000—the Pew Commission—that is, the Pew Charitable Trusts in association with the Johns Hopkins University School of Public Health—issued a report severely critical of our national approach to chronic disease. The report concluded that this spending imbalance fostered the widespread failure of the current public health system to adequately address the role of the environment in the understanding and prevention of chronic illness. The Pew Commission findings reflect what the history of allergy in America reveals: we no longer can afford to neglect how environment and allergic disease interact.[14]

We must better understand how perceptions of environment and illness shape one another. A more integrated, ecological perspective offers hope toward this end; it offers the opportunity to recognize, through a wider view, instances of progress in reducing the incidences of allergy—as in the 1970s in New Orleans, when the replacement of dilapidated housing stock and improved access to health care for the city's poor resulted in a marked decrease in hospital admissions for asthma.

This book offers such an ecological look at the history of allergic disease—a panoramic view of the changing relationships between illness and environment in America. From its early appearance in the nineteenth century among America's urban elite, allergy has spread throughout North America, moving across boundaries of geography, race, and class. This spread has coincided with both changing causal explanations and changing environmental conditions. Environmental changes were driven, not only by burgeoning populations, the effects of agricultural and industrial progress, and a general westward population movement, but also by the very efforts to circumvent the disease. The migration of people to areas free of hay fever was often followed by the influx of hay-fever-producing plants like ragweed, which thrived in the disturbed soils created by human settlement. The spraying of pesticides in public housing to combat infestations of cockroaches, a known asthma trigger, has exposed children of the urban poor to toxic chemicals and largely unspecified dangers. History shows that by steadfastly ignoring the complexity of environmental interactions in the search for simple solutions, we have helped to create America's allergic landscape.

CHAPTER

1

Hay Fever Holiday

Hay Fever to-day is an American specialty. . . . In no other country does Hay Fever give so much employment or cause so much prosperity.

—William Hard, 1911

August was Edwin Atkins's least favorite time of year; it was when his annual debilitating cold always appeared. But in the summer of 1869, when he was nineteen years old, the young Mr. Atkins tried something different. Some years earlier, Edwin's father, Elisha, had built a summer estate ten miles outside Boston in the newly fashionable suburb of Belmont, far removed from the stresses of city life and the mental strain of the family's thriving merchant business in the Cuban sugar and molasses trade. But the house on Wellington Hill, overlooking the nearby orchards and farms, offered Edwin little relief from the malady that was sure to arrive as the apples, pears, and other fruits ripened in the hot sun. Two months that Edwin spent largely prostrate indoors, his body overcome by a deluge of tears, throat irritations, limpid nasal fluids, and coughing spells, was more than he cared to endure again.

This year, wealth combined with yet another advantage to offer Edwin hope of escape. Edwin's father had recently been elected to the board of directors of the Union Pacific Railroad, which just three months earlier had joined its tracks with those of the Central Pacific to unite the United States' eastern and western shores. On 13 August 1869, Edwin left Boston on a Union Pacific train bound for Sacramento. Riding on the rails of a company his father helped finance and in the luxurious comfort of a Pullman car, Edwin was leaving his family, and with any luck his annual affliction, behind. As the train sped across Pennsylvania into the western parts of Indiana and on to Iowa, Edwin reported in his travel journal a slight cold in the head and throat, a typical early sign of the seasonal paroxysms and misery that would normally follow. But as the train reached the uncultivated prairies of Nebraska, the illness seemed to leave his body. He remained quite well during his two-month sojourn, at least until the journey home. As he crossed the Mississippi River on 22 September, the "unmistakable symptoms of [his] Autumnal Catarrh appeared."[1]

Catarrh, an inflammation of the nose and throat with an increased production of mucus, can be the common cold. But autumnal catarrh is a seasonal affliction and can also be, as Edwin discovered, subject to regional effects. Endemic to some regions, absent in others, the disease that plagued Edwin each summer and fall appeared mysteriously linked to geography and place. To the average tourist speeding across the continent, changes in altitude and temperature, soil type and vegetation, or sunlight and dust were of little consequence. But to Edwin, they were prominent features of the landscape and somehow tied to his health. He was, in the words of the nineteenth-century Boston physician Morrill Wyman, a catarrhoscope, with a body so sensitive as to detect even slight seasonal—and, as Edwin discovered, locational—changes.[2] Edwin didn't know exactly why his body reacted so vehemently when he crossed the Mississippi. But he did know that if he continued traveling directly east, the symptoms of his catarrh would become more severe. He changed his route, booking a ticket on the Great Western Railway of Canada, and headed north in hopes of skirting around the region where he knew sickness would prevail. The invigorating air of the northern Great Lakes region offered the palliative Edwin sought. He arrived in Boston on 25 September, having suffered but one sharp attack of the catarrh. All in all, it had been a most successful hay fever holiday.

Edwin's travels were the beginning of a fashionable trend. By the 1880s, hay fever had become the pride of America's leisure class and the basis for a substantial tourist economy that catered to a culture of escape. For example, the Reverend Henry Ward Beecher, "archbishop of American Liberal Protestantism"—whom Sinclair Lewis once described as a combination of St. Augustine, Barnum, and John Barrymore—found the White Mountains of New Hampshire to be an ideal retreat. His summer cold plagued him each year with such regularity that he was able to time his six-week vacation away from Brooklyn and his congregation at Plymouth Church by it.[3] "I have never spent a summer in the city, and shall never, if I can help myself," boasted Beecher. "I had rather have 'Hay Fever,'" he said, finding the illness a convenient excuse for his need to escape. Another sufferer, Henry W. King, Chicago's most prominent wholesale and retail clothing merchant, found Mackinac Island, located in the straits where Lake Michigan and Lake Huron meet, a "delightful place of summer resort." It also offered immunity from the "peculiar disease" that visited him each August.[4]

Seeking refuge from the watery eyes, flowing nose, sneezing fits, and asthma attacks that appeared with the "regularity of a previously calculated eclipse," these "accomplished tourists" also sought a holiday from the "desk, the pulpit, and the counting room" of the city.[5] Many physicians believed a nervous predisposition to be a necessary precondition for the development of the ailment, and it was thought that in urban spaces, just such a nervous tendency prevailed. Hay fever, in the opinion of American physician George Beard, was a functional nervous disease that bore a close relation to the much-celebrated American malady Beard did much to promote in the late nineteenth century: neurasthenia—that is, nervous exhaustion. In his widely popular 1881 book, *American Nervousness,* Beard pointed to modern civilization, and particularly American civilization, as the source of nervous exhaustion, which included among its many symptoms sensitivity to climatic change and "special idiosyncrasies in regard to food, medicines, and external irritants." An extremely sensitive nervous system, coupled with the depressing influences of heat, Beard believed, made a particular class of individuals susceptible during the dog days of summer to a host of external irritants, including dust, sunlight, and plant pollens. In the absence of effective drugs, removing oneself from the cause to a so-called exempt place became the preferred remedy among the country's afflicted bourgeoisie.[6]

Hay fever holidays enjoyed by America's well-to-do were part of an expanding nineteenth-century tourist trade. Leisure had become both a popular pastime and a marketable commodity after the Civil War. And, as noted in the introduction, hay fever began as an illness that only the wealthy could afford to treat. In the White Mountains of New Hampshire and along the northern shores of Lake Michigan, as well as many other places, hay fever catalyzed a lucrative recreational and tourist industry patronized largely by wealthy urbanites. In leisure and nature, not to mention the pleasantries of fine society on holiday, they sought an antidote to the hustle and bustle of the city that left their minds and bodies fatigued and susceptible to this modern malady. By the late nineteenth century, the forests of Michigan, New Hampshire, and other regions had been heavily logged and were largely exhausted. It would take a disease to bring new life to these worn lands. Hay fever tourism not only reinvigorated weary bodies and landscapes but also reinvented nature into an economic resource for health and pleasure. Like axe and plow, hay fever left a mark upon the landscape, visible to this day.

Hay fever first became medically prominent, not in America, but in Great Britain. In 1819, Dr. John Bostock delivered a paper to the London Medico-Chirurgical Society, describing his own condition of a catarrhal inflammation of the eyes and chest that appeared regularly each year during the early summer season. After reporting on twenty-eight additional cases and spending nine more miserable summers, Bostock in 1828 named the disease "Catarrhus Aestivus," or "Summer Catarrh." Other physicians in England, France, Germany, and Switzerland began to report similar cases, in which they or their patients suffered symptoms resembling those of catarrh or the more debilitating effects of wheezing associated with asthma during particular times of year. By the 1860s, "hay fever" or "hay asthma" had become a commonplace term used to describe the recurrent symptoms of swollen eyes, frequent sneezing, and laborious breathing most noticeable in England and other European countries during the hay-making season.[7]

The type of individual affected by this strange new malady was anything but ordinary. The Manchester physician Charles H. Blackley observed in his seminal 1873 treatise on hay fever that among his hay fever patients the overwhelming majority were clergy and doctors. Many physicians shared Blackley's opinion that hay fever was an "aristocratic disease"; if it was not "almost wholly confined to the upper classes of society, it was rarely, if ever, met with but among the educated."[8] Nineteenth-century doctors theorized that among the wealthy and educated, a nervous temperament prevailed, "fostered and perpetuated with the progress of civilization and with the advance of culture and refinement." Its effects could be observed especially among the "brain-working population" in a host of disorders including dyspepsia, neuralgia, insomnia, and nervous exhaustion.[9] Hay fever was yet another "condition of the nervous system which mental training generates."[10]

Why had hay fever, unknown prior to the nineteenth century, become more common? Blackley turned to the economic and environmental history of Great Britain to account for its apparent rise among urban professionals. Before the Industrial Revolution, he argued, a large portion of the population in England was exposed to the atmospheric conditions of country life, either through the cultivation of the soil or the production of woolen, linen, and cotton goods, largely in rural villages and towns. As England's population increased, large numbers of people moved from the "country to the workshops and mills of towns." In doing so, they removed themselves from pollen and other exacerbating factors to which agricultural laborers were continually exposed. At the same time, the influx of population into the cities, where greater educational opportunities, wealth, and luxury prevailed, created circumstances "favorable to the development of the pre-disposition to hay fever." The frenzied pace of urban life, the mental demands of modern business, and the removal from nature, which could fortify the body and calm the hurried mind, had strained the nervous systems of the city's educated and well-to-do classes. "As population increases and as civilization and education advance," Blackley warned, hay fever "will become more common."[11]

In the United States too the increasing differentiation—physical and social—between urban and rural life created conditions in which hay fever flourished. Morrell Mackenzie, a physician at the London Hospital in the late 1800s, may have prided himself that the "national proclivity to hay fever" in Britain offered "proof of our superiority to other races."[12] But by the 1870s physicians and sufferers in the United States were making similar claims. Hay fever, political journalist William Hard boasted, had become an "American specialty . . . the English compete with us no longer." "In no other country are summer resorts built up on Hay Fever patronage," remarked this hay fever sufferer and friend of Theodore Roosevelt. "In no other country is the Hay Fever travel toward certain regions so thick that railways serving those regions might well enter Hay Fever with the Interstate Commerce Commission as the basis for part of their capitalization. In no other country does Hay Fever give so much employment or cause so much prosperity. It has come to deserve to be a plank in the national platform of the Republican party."[13]

Like many hay fever sufferers, Hard employed humor and hyperbole to great effect. But his claims were not completely facetious. Although hay fever never became the center of Republican politics, it did find a place in Whig Party affairs. Daniel Webster, prominent Massachusetts senator, Whig Party leader, and twice secretary of state, was the most celebrated hay fever sufferer in nineteenth-century America. Webster dated his first attack to 1832, when he was fifty years old. His annual cold commenced about 23 August, accompanied by fits of sneezing and profuse discharges from the nose. On 15 August 1849, he wrote, "In seven days I shall begin to sneeze and blow my nose; and the first week the catarrh is usually most severe." His eyes became progressively swollen, preventing him from reading and limiting the stroke of his pen to signatures. By the middle of September, Webster's disease would move into its last recognizable stage—asthmatic. In the fall of 1850, Webster wrote to President Millard Fillmore that given the "annual occurrence of his illness" and his long absences from Washington, perhaps he ought to consider himself unfit for the holding of public office. Two years later, Webster resigned his post as secretary of state for health reasons.[14]

One of the nation's most highly paid lawyers, branded by his political adversaries as a friend of the rich, Webster adopted a lifestyle in keeping with those of his clients and political supporters. Hay fever was another bond that linked him to America's leisure class. Like his British counterparts, Webster found the "bracing air of the ocean beneficial" in offering at least partial respite from his annual symptoms.[15] Fleeing the heat and mental strain of the nation's capital, Webster often retired in summer to his eighteen-hundred-acre coastal estate in Marshfield, Massachusetts, eleven miles north of Plymouth. The severity of the hay fever season at Webster's farm depended greatly upon the weather. A northwesterly wind blowing across the land and out to sea spelled misery. But a day on the ocean aboard one of his seven yachts, when he fished for cod or haddock or shot coots near shore, replenished Webster's worn nerves and fortified his body against an attack that would leave him depressed and fatigued. Webster's friend Samuel Lyman remarked how a morning of outdoor sport at the Marshfield retreat did more to "repair the inroads upon one's health, made by too much application to books, business, or mental labor . . . than by the idle monotony of a dozen days spent at Saratoga, or any other mere watering place." According to Lyman, the rigors of outdoor life, for which Webster, who had grown up on a New Hampshire farm, had acquired a fondness, had transformed the once frail and sickly child into a man with a robust constitution. Not surprisingly, Webster turned to outdoor recreation in nature as a tonic to strengthen him against his enfeebling seasonal disease.[16]

Still, relief was never complete. In 1849, Webster experienced a much milder attack of his annual malady while visiting his boyhood farm in Franklin, New Hampshire, forty miles south of the White Mountains. Two years later, Webster returned to the family farm in early August to conduct some business. Inundated by visitors and knowing the catarrh would soon be upon him, Webster journeyed to the White Mountains, seeking quiet, privacy, and solace. "Thinking that the mountain air might [give him] general health and strength [to] at least in some measure resist (the catarrh's) influence and mitigate its evils," Webster waited for his "enemy . . . to attack." On 19 August he wrote in expectation: "Four days hence is the time of its customary approach." The days of 23, 25, and 27 August passed, and still there were no signs of catarrh. Webster mostly escaped his dreaded ailment that year, though it did appear on 8 September, when the train in which he was riding was derailed in Boston en route to his Marshfield residence. It was nonetheless a remarkable success, which he attributed not to his surroundings but to a regimen of iron, potash, and arsenic, recommended by a sufferer of June cold. To the many pilgrims who later journeyed to the White Mountains, however, Webster's account became hallowed testimony and he, a witness to the therapeutic powers of place. Webster's brief mountain sojourn became a part of the memory and landscape of the White Mountain region, which grew to become America's most luxurious and popular hay fever resort.[17]

Hay feverites, visiting the White Mountains in increasing numbers after the Civil War, drew upon and contributed to a growing tourist industry that catered to an educated elite largely centered in the urban and industrial East. Webster had first visited the region in 1831, when the area was frequented by a group of distinguished artists, writers, and scientists, including Thomas Cole, Ralph Waldo Emerson, Nathaniel Hawthorne, Benjamin Silliman, and Henry David Thoreau. These men were able to capitalize on the White Mountain scenery in promoting their own careers. As a region on the economic margins of New England, the White Mountains, with thin, rocky soil that was unable to sustain farming as a livelihood, appeared a desolate and primitive landscape to early nineteenth-century New Englanders. More civilized landscapes were the preferred tourist attractions on the American Grand Tour, the standard itinerary of America's well-to-do travelers. Such places included the Connecticut and Hudson River Valleys, which could promise panoramic views complete with "prosperous farms, fine homes, and thriving villages, as well as dramatic mountain views."[18] In the early 1800s, Ethan Crawford's main clientele at his small farm and inn near the Notch of the White Mountains were farmers and merchants traveling through the mountain pass on the road between Portland, Maine, and New Hampshire's northern interior. In the 1820s, however, wealthy and educated travelers from eastern cities slowly began to appear, many of them intent upon climbing Mount Washington.

In August 1826, a day of steady rains following a summer drought triggered a series of landslides that altered the course of the Saco River. The events destroyed the turnpike that ran through the Notch. Miraculously, the Samuel Willey house, the homestead nearest to Crawford's, was protected from a large mudslide by a boulder behind the house that diverted the slide to both sides. But tragically all nine of the Willey household, including five children, perished as they fled outside, presumably upon hearing the mountain roaring down upon them. The Willey disaster became the subject of extensive press coverage, paintings, and literary and philosophical speculation and helped to transform the region into a tourist attraction overnight. For New Englanders, the incident presented an unfathomable mystery. The Willey house became, like the ruins of ancient civilizations in the Old World, a reminder of man's mortality and of God's untold power. By the 1850s, nine hotels in the immediate vicinity of Mount Washington and several more in the surrounding area provided accommodations for as many as five thousand visitors. Many who came visited the Willey house and climbed Mount Washington, expecting to imbibe the moral sublime. Timber had supported a large potash industry in the White Mountains, but wild and scenic nature became the resource that initiated and sustained a booming tourist economy.[19]

God and nature, it must have seemed, dwelled deep in this wild mountain region, and New England transcendentalists increasingly came to revel in its glory. Crawford was one of the first in the area to cash in on the passion for nature cultivated among the educated elite of Boston, Cambridge, Concord, and other New England towns. He began to transform the land to accommodate the interests of this new clientele. By 1831, he was able to accompany Webster to the summit of Mount Washington on a path he had spent years clearing and improving to enable easy access by foot or horseback.

The patronage of wealthy hay feverites added significantly to the economic boom. Sufferers fleeing Boston could reach the area on the Grand Trunk Railroad in less than ten hours; those escaping New York City could make the trip in less than twenty-seven. The small village of Bethlehem, New Hampshire, located fifteen miles northwest of the Notch, became the mecca of an annual pilgrimage.[20] When Helen Hunt, whom Ralph Waldo Emerson would name America's greatest poet, visited the area in September 1865, she described Bethlehem as a "place not yet ready for strangers, but [a place] meant to be."[21] Her words were prophetic. In a matter of twenty years, the town was transformed from a hamlet with one small inn into a thriving resort with thirty boarding houses and hotels and as many as five hundred seasonal hay fever residents. By the early 1900s, the number of hay fever victims visiting Bethlehem during the summer months reached two thousand.[22]

Although the White Mountains area was not the only exempt region known to sufferers, it was the first to capitalize on its natural resources in developing a sizable industry that catered to the hay fever tourist trade. Morrill Wyman, who suffered his first attack of hay fever upon his graduation from Harvard in 1833, began his annual summer pilgrimage to the White Mountains in the 1860s, after hearing of the beneficial properties from a fellow sufferer and patient. Wyman's own family offered abundant material for studying the natural history of the disease: his father, two brothers, sister, and two children were all sufferers. By the time he published his influential book, *Autumnal Catarrh,* in 1872, Wyman had extended the study of his own family to include eighty-one case histories of the disease; fifty-five of these sufferers reported either partial or complete relief while residing in the White Mountains. Wyman gathered many of these cases by walking through the region, inquiring at hotels and inns about guests who had come in search of hay fever relief.[23] Wyman differentiated between two annually recurring catarrhs. One, summer catarrh, first described by Bostock and more commonly known as hay fever, June cold, or rose cold, appeared in the last week of May or first week of June and lasted roughly four or five weeks. The other Wyman named autumnal catarrh. Unknown in England but the source of

Daniel Webster's misery, it commenced the last week of August and persisted until the first frost. Removal to a noncatarrhal region offered the only known cure (Figure 2).

Through word of mouth and the advice of physicians like Wyman, news of the healing properties of the White Mountains spread. In the region, and in Bethlehem in particular, "the climate, the soil, the hotels, and the railways [seemed] arranged by a special kind of providence to ameliorate [the] sufferings of the hay-fever patient."[24] On 15 September 1874, hay fever sufferers gathered to establish the U.S. Hay Fever Association in Bethlehem, where it continued meeting for the next fifty years. "Among all the maladies of men," reflected the retiring president, Samuel Lockwood, in 1891, "this is the only one which has crystallized into an organic companionship. It came into being as the result of two forces—the attraction of the social magnet, and the gravitation to a common center—the hope that in the social impact of suffering and inquiry might be found the secret of relief, and perchance complete deliverance."[25]

Unlike the Adirondacks, the Rocky Mountains, and southern California, the White Mountains region did not market its climate and locale to a wide range of pulmonary sufferers, most notably those stricken by tuberculosis. Neither did Petoskey, Michigan, where the Western Hay Fever Association was established in 1882. Both the U.S. Hay Fever Association and the Western Hay Fever Association were fashionable societies that claimed a disease and a region as signs of exclusiveness. As one hay fever skeptic remarked, "People who have small-pox or scarlet fever, or even gout, have never formed a small-pox club, a scarlet fever society, or a gouty men's association." But in the case of hay fever, being seen was precisely the point. As a disease of the genteel elite, hay fever became a part of the conspicuous leisure and consumption that characterized Gilded Age resorts.[26]

The witty speeches delivered by prominent members of the U.S. Hay Fever Association, such as the Reverend Henry Ward Beecher—whose residence and Sabbath sermons at Twin Mountain House, a Mount Washington area grand hotel, were built-in advertisements for the area—gave the society a jocose flair in its early years and made great newspaper copy. Beecher remarked at an association meeting in 1879 how he disliked the "ingratitude . . . shown by hay fever patients who were constantly maligning a disease which sticketh closer than a brother. . . . For his part, he could never be grateful enough for having been thought worthy of

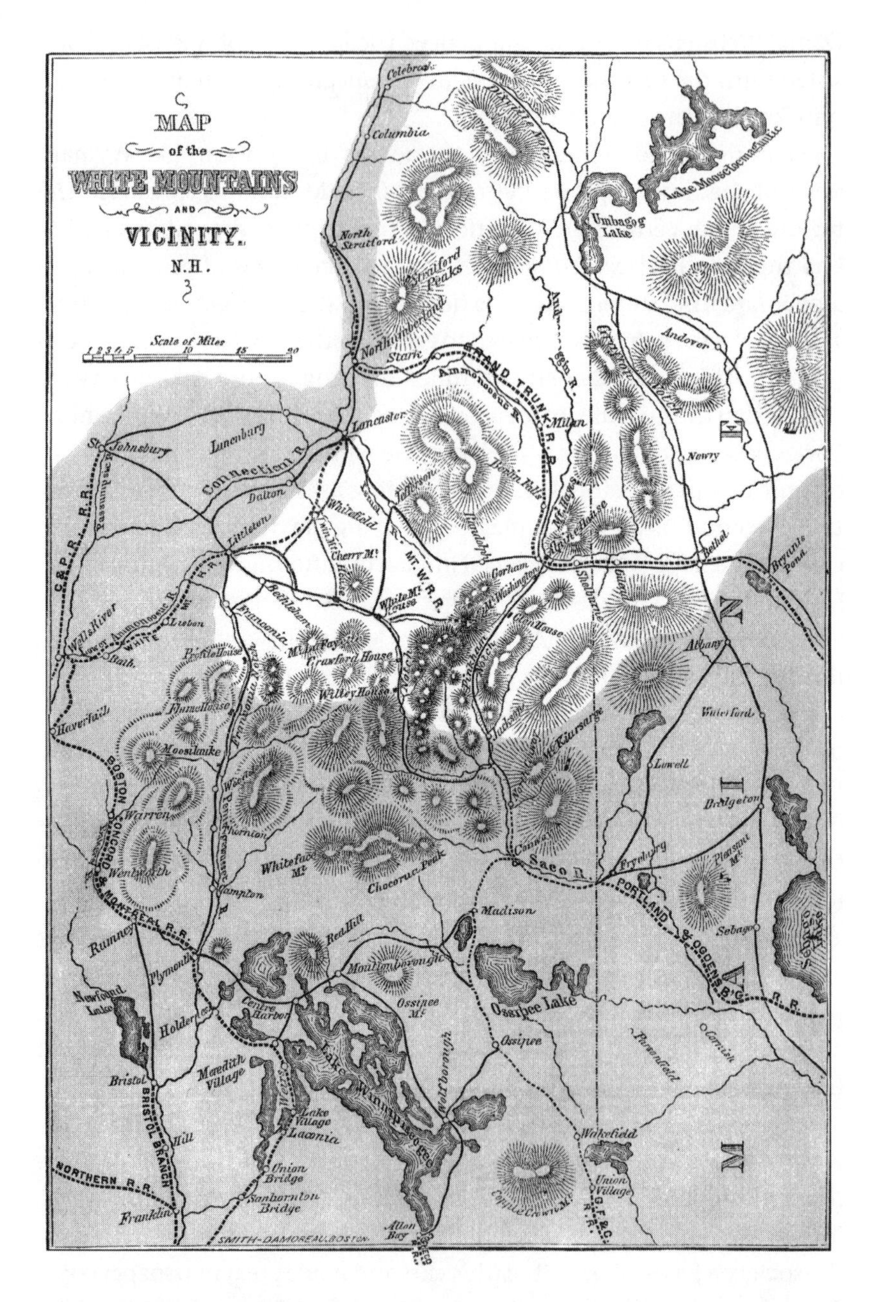

Figure 2. Morrill Wyman's map of the White Mountains. The gray shaded areas offered no refuge; only in the white regions did hay feverites find escape from their dreaded disease. Morrill Wyman, *Autumnal Catarrh* (New York: Hurd and Houghton, 1872), frontispiece.

enrollment in the ranks of Hay Feverites." "I can't get well, and it seems to me that I won't get well," quipped Beecher. "I esteem my six weeks' vacation in the mountains too well."[27]

Red-nosed humor reflected a certain class aesthetic of hay fever as a disease, just as pallor and sadness accompanied mid-nineteenth-century middle- and upper-class romantic notions of consumption, as tuberculosis was then called.[28] Humor spoke tellingly of class membership in the U.S. Hay Fever Association (Figure 3). Although anyone who was afflicted with hay fever or rose cold and who paid $1 could join, the association overwhelmingly comprised doctors, judges, lawyers, ministers, merchants, and other educated male professionals. For these members, a six-week vacation away from business and family was little problem, but the same holiday would cost a laborer between six and twelve months of wages. For those who couldn't afford the expense of escape, hay fever was not so humorous a disease. "Unlike most hay-fever victims, I am neither rich nor intellectual, and must stay at home," wrote a sufferer to the association's president. "Up to the day I take the fever, I love my life. Afterward, I often pray that I may die."[29] The experience of the less fortunate, forced largely to remain at home, was shaped by medications, patent nostrums, and attempted cures, not by enjoyable holidays. Lest the less fortunate be left with no aid, the U.S. Hay Fever Association published in its annual manual a list of remedies that individual members had found useful for relief. Galvano-cauterization, cocaine, hydrozone (ozonated water), and the ice bag treatment were just a few of the remedies that came in and out of vogue through the years.[30]

Hay fever resorts reinforced the impression that hay fever was a white, middle- and upper-class disease, despite letters from occasional less fortunate sufferers that suggested otherwise. E. J. Marsh, a physician in Paterson, New Jersey, recognized this problem in the hay fever censuses conducted by both Wyman and Beard, who gathered their disease statistics almost solely in hay fever resort areas. Not surprisingly, they found that the disease allegedly afflicted twice as many men as women and that it was rarely found among African Americans or Native Americans. In contrast, when Marsh undertook a survey in Paterson, he found hay fever to be as common among manual laborers as white-collar professionals. The presumption that hay fever afflicted only wealthy white individuals was shaped largely by the exclusiveness of hay fever resorts

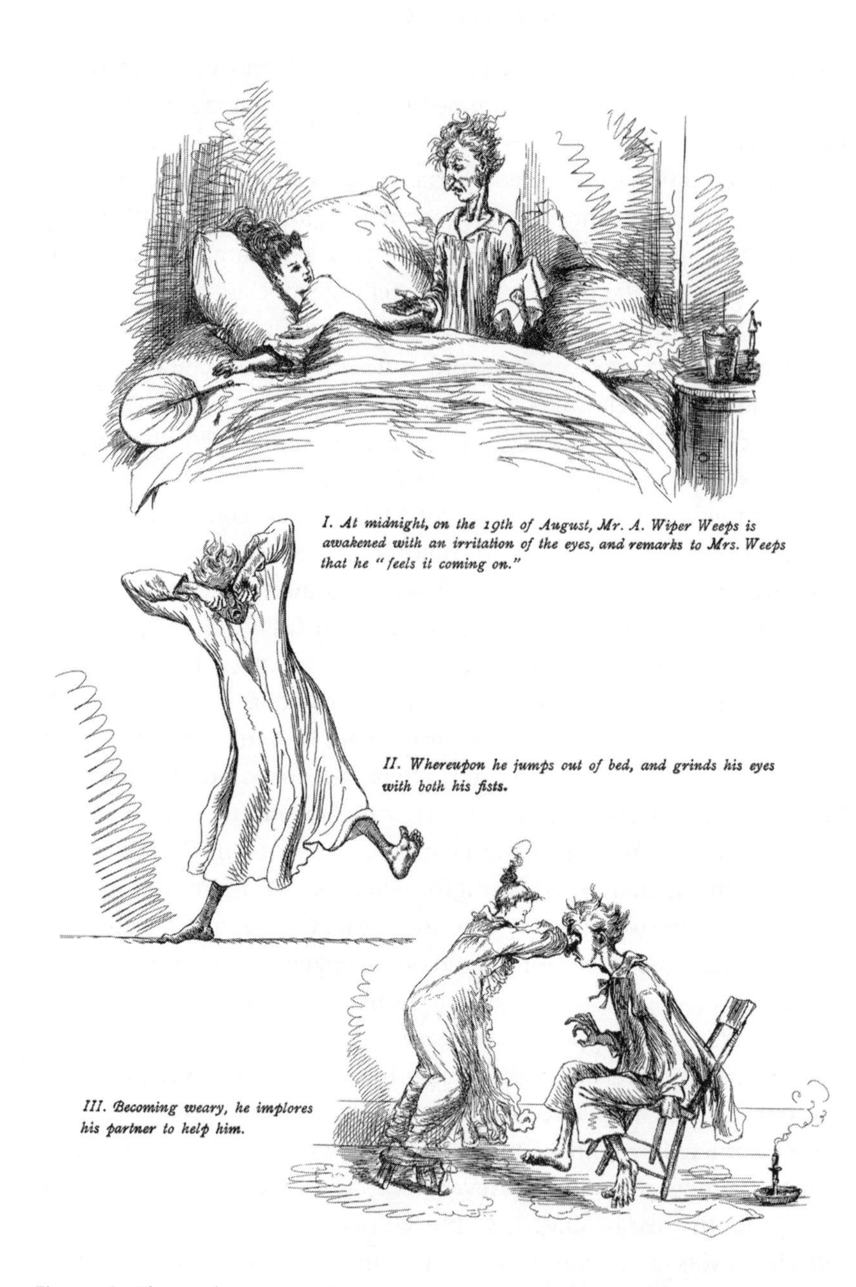

Figure 3. The early stages of hay fever, illustrated through the adventures of Mr. A. Wiper Weeps. Humor spoke tellingly of the class identity of hay fever in nineteenth-century America. Courtesy of Ebling Library, Rare Books and Special Collections, University of Wisconsin–Madison.

and the geographic differentiation of health care in American society along race and class.[31]

Although the "mysterious bond in their throats, . . . noses, and eyes," made hay fever sufferers, in the opinion of one writer, "first cousins, if not brethren," social status limited how far such a brotherhood could extend.[32] George Scott, a hay fever exile from the Midwest, may have rejoiced in his infirmity, believing that it offered him the promise of social mobility. But contrary to Scott's hope that he be "classed among the intellectual giants of America," hay fever did not open the doors to privilege for those on the lower rungs of the social ladder.[33] Perhaps a moral predisposition or the spirited sermons of Beecher inspired a few to make charitable appeals on behalf of the less fortunate sufferers. But appeals to the Protestant social gospel fell on deaf, and sometimes hostile, ears. In 1889, when a Boston woman urged the U.S. Hay Fever Association to launch a fund-raising effort for the establishment of a home "in the mountains for poor people unable to visit them," she was met with sharp reproach. Colonel M. Richard Muckle, the association's president and business manager of Philadelphia's popular newspaper, the *Public Ledger,* "was satisfied that $1,000,000 would not be sufficient for that purpose." The association, he reminded her, "was endeavoring to find a remedy for Hay Fever, which would enable poor people to stay comfortably at home."[34]

Hay fever exiles looked suspiciously upon almost all proposed cures, as they did upon sufferers who could not afford the holiday cure. Meetings of the association, where personal experiences were recounted, also served as a forum for individual testimonies about remedies and exempt places that had been tried. Although a few esteemed brethren claimed to be cured, they inevitably found their way back to Bethlehem. Hotel proprietors may have looked anxiously upon curative treatments, since an estimated one-quarter of their summer customers were hay fever sufferers, but they readily consoled themselves each year as the U.S. Hay Fever Association proclaimed Bethlehem and other exempt places to be the only effective remedy.[35] Such proclamations were essential to the identity of hay fever exiles, for they were bound to the lifestyle of leisure that their disease both afforded and, in their opinion, necessitated.

The grand hotels built after the Civil War to accommodate wealthy tourists furthered the making of hay fever into a fashionable disease of prosperous white society. Able to house five hundred and three hundred guests respectively, the Maplewood Hotel and Sinclair House (the latter a favored meeting place of the U.S. Hay Fever Association) lavished luxury and culture upon their guests at rates almost double those of New York City's first-class hotels (Figure 4).[36] Karl Abbott recalled the highlight of summers in Bethlehem, when the Maplewood hosted the Grand Ball and Cotillion. "Up the splendid stairway and into the mammoth high-ceilinged ballroom swept people whose names made the society columns in New York, Boston, and the southern and European spas," he wrote.[37] There a twenty-piece orchestra of Boston Symphony musicians met them. In these resorts, culture and nature had been carefully refined. To the "pleasing aspect of finish and cultivation" that adorned these grand hotels, proprietors improved upon the "entrances of shady paths or rocky hills" through which guests passed to "woo the goddess Nature."[38]

In fashioning a landscape of leisure, hotel proprietors and their guests in the White Mountains understood cultivation in terms of art, not agriculture. Nineteenth-century philosophers like Ralph Waldo Emerson and landscape architects like Frederick Law Olmsted believed humans could improve upon and purify nature through art.[39] If the "rugged grandeur" of the mountains was "somewhat marred by the presence of [such] mammoth hostelries, and by the lines of railroad which pass with sinuous course through the valleys, . . . far better this," cried the editor of the region's summer weekly, the *White Mountain Echo,* "than the time when only a few could visit them in their primitive grandeur." Within fifty years of Daniel Webster's first visit to the region, "a few small farm houses and poorly cultivated fields had given way to large and tasteful residences."[40]

The estates and grand hotels of the White Mountains that had sprung up after the Civil War were populated largely by the middle- and upper-class residents of large eastern cities who had "wrung out of the nervous hand of commerce enough means to realize" an extended summer holiday.[41] They drew poetic inspiration from, botanized, and took solace in nature; they did not care to toil in it. They also did not care to see nature disturbed, for example, by agricultural practice or logging. The regional transformation of the White Mountains into an idyllic retreat depended upon distancing the forces of production, of not only the factory but also the farm. Trains carried urbanites into the valleys of the White Mountains. But they also carried Midwest grain to the

Figure 4. Exterior and dining room of the Maplewood Hotel, ca. 1890. Grand hotels like this one helped transform the White Mountains by the late 1870s into a genteel tourist region that catered to a northeast urban professional clientele. Courtesy of Bethlehem, New Hampshire, and Bondcliff Books.

east, further diminishing the importance of agriculture to the White Mountain region's economy. In bringing the hinterlands closer to the city and the city closer to the White Mountains, the railroads helped transform the White Mountains into a genteel tourist region that drew upon Romantic ideals and urban income to fashion the area as cultivated wilderness.[42]

Thus the community that hay fever exiles established in the White Mountain region was held together by disease, class, and place. By the 1880s, the disease had become as prominent a seasonal feature in the society pages of urban newspapers as the cotillions of resort towns. One skeptic thought "the unfailing certainty with which hay fever patients select the most attractive and fashionable Summer resorts" was a sure sign that hay fever was "simply the creation of hotel-keepers."[43] Sufferers, of course, disagreed. In the White Mountains they found physical relief, as well as a place that affirmed their social identity and shaped their relationship to the natural environment. Hay feverites became active agents in shaping not only the physical landscape of the White Mountains but also what can be called its geography of place—that is, the area's larger cultural meaning and value: the material relationships of daily life, the social contours of the region, and the symbolic space that nature inhabited. Hay fever became the malady on which White Mountains summer residents prided themselves, even as they escaped the nervous energies of the city to breathe more deeply and freely in a refined nature that sustained body, soul, and a thriving tourist trade.

If civilization and progress spawned hay fever, then there was perhaps no better breeding ground than the burgeoning metropolis of Chicago. No other city in nineteenth-century America had risen so quickly to rival established eastern centers of wealth and power. In less than fifty years, this former military and trading outpost had become an empire of commerce and industry. By the 1890s, only New York City surpassed the reach and extent of the markets and capital concentrated in Chicago.[44] When Congress in 1893 sought a site to commemorate the four hundredth anniversary of Columbus's landing in North America, it passed over New York City and chose Chicago as the place to create an edifice of cultural and industrial achievement that would rival the Eiffel Tower, which had been built four years earlier for the 1889 Paris International Exposition. More than 27 million adults and children came to the World's

Columbian Exposition to witness the splendor of the White City erected on the shores of Lake Michigan. Visitors in this "neoclassical wonderland" of ivory palaces and amethystine lakes "ablaze with countless lights," walking through temples of agriculture, machinery, and electricity, found themselves guided through the "steps of progress of civilization and its arts in successive centuries, and in all lands up to the present time." It was a dizzying display of modern progress that could easily overwhelm the mind and the senses. Horace Benson, a Denver attorney, reported being so bedazzled by his visit to the exposition that he returned to his room, "tired in mind, eyes, ears, and body" and fell into a chair "completely exhausted."[45]

Progress could have that effect on the body, as nineteenth-century hay fever sufferers knew only too well. It was why Clarence Edmonds Hemingway left his medical practice in Chicago each summer to go to the family cottage on Walloon Lake, eight miles south of the small town of Petoskey, Michigan, at the head of Little Traverse Bay on the eastern shore of Lake Michigan. Hemingway kept a hurried schedule as a physician, attending to the needs of his private patients and those of the Oak Park hospital while also serving as medical examiner for the Borden Milk Company and three insurance firms. Six feet tall, barrel-chested, and bearded, the doctor displayed the kind of nervous temperament, intellectual wit, and success that were the trademark of many nineteenth-century hay feverites. He also had a passion for manly outdoor recreation—hunting and fishing in the forests, streams, and lakes of northern Michigan—activities that he believed strengthened his constitution against the enervating conditions of urban life. Frank Lloyd Wright once suggested that Oak Park residents, like Hemingway and his wife Grace Hall, "took asylum there to bring up their children in comparative peace, safe from the poisons of the great city."[46] But the genteel Chicago suburb was not far enough removed from the metropolis to free Hemingway of civilization's disease. A more remote location needed to be found.

In 1899, Hemingway, his wife, their one-year-old daughter Marcelline, and their newborn son Ernest journeyed for the first time to their upper Michigan lakefront retreat. At the time, the land and waters that inspired Ernest Hemingway's early short stories had already become the bases of a bustling hay fever tourist trade. Magnificent steamships like the *Manitou,* a "floating palace" 295 feet in length, cruised the waters of the Great Lakes, transporting families like the Hemingways

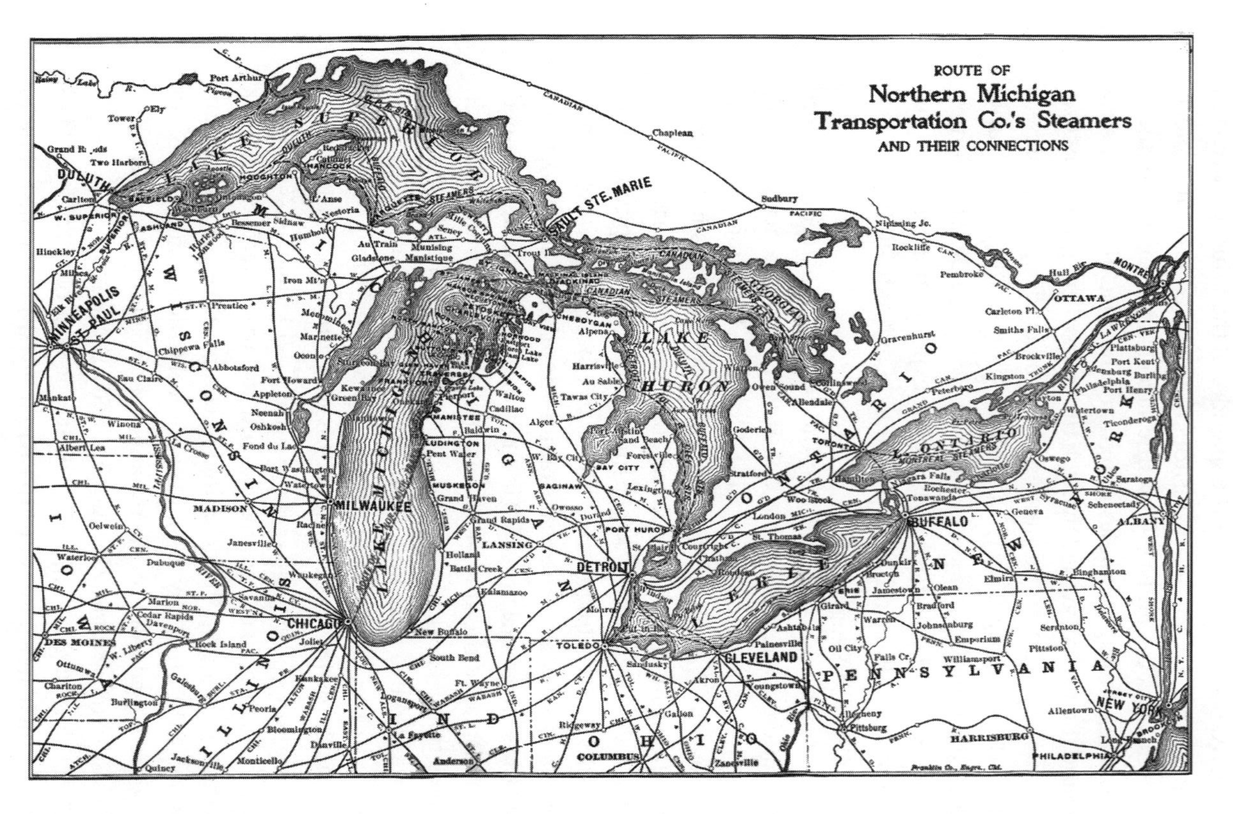

Figure 5. By the 1880s, steamships and railroads put northern Michigan's summer retreats within a few days' journey from Chicago, Detroit, Cleveland, and other Midwestern cities. The family of Clarence Edmonds Hemingway was among the many that summered in northern Michigan to escape the disease of civilization, hay fever. Courtesy of Mackinac State Historic Parks.

from Chicago to Charlevoix, Harbor Point, and Mackinac Island—the ports of northern Michigan's summer retreats—in less than twenty-four hours (Figure 5). At the depot of the Grand Rapids and Indiana Railroad in Petoskey, where the Hemingways caught a smaller train bound for Walloon Lake, as many as one thousand passengers disembarked daily during the height of the tourist season.[47] They found a town ready to greet them. The Arlington Hotel, built in 1882 for $60,000 and later expanded to include elevators, steam heat, a forty-by-seventy-foot dance hall, and a twenty-four-foot veranda, catered to the needs of seven hundred guests who could afford its rates of $5 per day. Only the Grand Hotel, forty miles to the north on Mackinac Island, rivaled it in size, opulence, and splendor. In Petoskey, twenty-one hotels, offering the most lavish or more modest accommodations and rates, competed for the business of the roughly forty thousand tourists that made the town a place of respite from late June until 1 October, when the last of the hay fever tourists left. In the off-season, just four thousand residents remained.[48]

When the Grand Rapids and Indiana Railroad first reached the hamlet of Petoskey on New Year's Eve of 1873, not even the railroad financiers imagined how this site of a Catholic mission and government Indian school would prosper. In little more than a decade, Petoskey would become the flourishing hub of a community of resort associations, summer cottages, and grand hotels built within a forty-mile radius for the health and pleasure of middle- and upper-class visitors. Forests of virgin white pine had lured the railroad into the uppermost reaches of Michigan. Speculators and settlers staked their futures on lumber; the supply seemed inexhaustible: 4.25 billion board feet came out of Michigan forests in 1890, surpassing the lumber production of all other states. But 1890 was a peak year; the next decade ushered in the end of Michigan's white pine era. In a futile attempt to retain their share of the national market, timber companies turned to harvesting the state's hardwoods. But other entrepreneurs saw the makings of wealth in upper Michigan's natural climate and balsamic aroma, which had, it seemed, the power to attract those with money.[49]

Petoskey, perched on a bluff on the southern shore of Little Traverse Bay, was in 1875 a town of 125 people, 3 hotels, 6 saloons, 1 bakery, and a blacksmith shop. But the little town was blessed by prevailing westerly and northwesterly winds.[50] The hundred miles of clear, cool Lake Michigan waters to the north and west of Petoskey had a cleansing effect on air, acting to rid it of pollen and other vegetable matter. Reverend C. I. Deyo of Oxford, Michigan, discovered the pure air of Little Traverse Bay shortly after the railroad arrived and acclaimed the region's "shores with the same rapture that filled the soul of the Israelites of old when they beheld the land flowing of milk and honey."[51] Others came, slowly increasing the ranks of Petoskey's "Old Guard of Sneezers." Dr. F. M. Pennebaker of Pleasant Hill, Kentucky, stumbled upon the therapeutic powers of Petoskey when it was still "just an ordinary Indian Village."[52] In 1882, approximately 150 hay fever sufferers, bound together by "the most sacred ties of mutual sympathy and mutual friendship," formally established the Western Hay Fever Association. By 1899, the organization proudly proclaimed an affiliation of 3,000 people.[53]

Like its eastern counterpart, the Western Hay Fever Association organized to "furnish an opportunity for an exchange of views and encourage scientific investigation into the cause of the disease, its cure and places of exemption." But its real goal was "to render the sojourn of those who [came to northern Michigan] as pleasant and profitable as possible."[54] After eight years of meetings, Mrs. Andrew Oliver, treasurer of the association, had tired of hearing about everyone's infirmities, especially after a visitor, alarmed at hearing about hay fever, asked her, "Is it ketchin?" Oliver believed the organization should pay more attention to the entertainment and social life of its members, who were, after all, strangers in the land. Musical and literary performances were added to the weekly association meetings, and attendance swelled as high as five hundred, as the hay fever exiles scattered throughout the area came together in education, culture, and friendship.[55]

Within a few short years, Petoskey had become the "harbor of refuge for all hay fever victims."[56] Its newspaper, the *Daily Resorter,* served as the news source, gossip column, advertising circular, and guide for seasonal hay fever residents. Hotels boasted of the unique topography and location that made their particular spot "superior to any other point for the cure" of hay fever.[57] Manufacturers of patent medicines, such as Cushman's Menthol Inhaler, used the newspaper's pages to bark their wares to asthma and hay fever sufferers in an early form of targeted marketing.[58] Entrepreneurs, like the famed "Indian Doctress" Princess Viroqua, a specialist in the treatment of hay fever, set up practice in Petoskey to aid its seasonal patients (Figure 6). Steamship companies advertised the advantages of a lake journey, as compared with dusty rail travel. Other ad copy

The Indian Doctress.

Treatment of Hay Fever a Specialty

All chronic diseases, all Female Weaknesses. Main office, Kalamazoo Mich. During the months of August and September will be located at Petoskey, in the cottage opposite the Cass House. Consultation Free. Consultation can be had by letter. Enclose stamp for reply. [76-e o d-34

Figure 6. The seasonal influx of hay fever sufferers to the western shores of Lake Michigan created a captive market for healers like Princess Viroqua, the "Indian Doctress," who set up shop in Petoskey during August and September, the height of the hay fever tourist season. *Daily Resorter,* 31 August 1887, 4.

described the area as a place where "the towns are so situated that the balmy, purifying air from the limitless forests of pine blows over the tempering lake waters from every direction" and where "hay fever is not only unknown but eradicated from the system."[59] But none of these efforts matched the marketing campaigns and investments of the Grand Rapids and Indiana Railroad in selling nature's recreational and healthful benefits as the great harvest of Michigan's northern forests drew to a close.

Known in its day as "the Fishing Line" and the "Hay Fever Express," the Grand Rapids and Indiana Railroad was the main overland transportation route along the northwestern shore of Lake Michigan. Committed to making the region a tourist playground, it marketed its route to wealthy anglers, hunters, and hay fever sufferers. Often these customers were one and the same. In little more than a decade after the Civil War, four outdoor sporting magazines—*American Sportsmen, Forest and Stream, Field and Stream,* and *American Angler*—burst onto the national scene and spoke to the increasing popularity of outdoor recreation as a pastime among America's growing number of wealthy gentlemen, who also happened to be the largest class of hay fever sufferers.[60]

As noted, one such gentleman enthusiast of the outdoor life was the busy Dr. Hemingway. Ed Hemingway thought hunting wild game, plying streams and lakes with rod and reel, and cooking over an open fire under a hemlock canopy—skills that he bestowed upon his son Ernest—taught physical courage and endurance and how to cope with hardship. Particularly sensitive in the summer to irritants around him, Ed Hemingway also believed that such activities strengthened the moral and nervous fibers frayed by city life. How readily he would have identified with an article published in *American Angler* in 1886, written by Milwaukee physician Arthur Holbrook, that informed readers of the best hay fever and fishing vacation spots northern Wisconsin and Michigan had to offer, as discovered by the Gitche Gùmee Camping Club in a decade-long search for health and recreation.

Like Hemingway, Holbrook found no better time to enjoy the great outdoors than the hay fever season, which commenced in mid-August and lasted into October. It was a time "when the woods are in their loveliest dress and camping can be enjoyed with the least fear of exposure or the least dread of mosquitoes and insects." Hotels were no longer crowded. And in the northern Great Lakes, the trout fishing season extended to 15 September. "To do something new and different from the ordinary daily habit at home, to live for a time away from the atmosphere of business, fashion and society, and to quietly and contentedly breathe the pure, wholesome, and inspiring air of nature and solitude," Holbrook assured his readers, gave "new life and strength to the body and soul."[61] Readers of Holbrook's article—primarily well-to-do male professionals from Chicago, Cincinnati, Detroit, Kansas City, Indianapolis, and other Midwestern towns who had disposable income, sniffling noses, and time to spare—were precisely the customers the Grand Rapids and Indiana Railroad hoped to lure north. They came to nature in search of renewed health of mind and body.

The search for manhood and health attracted many who boarded the Grand Rapids and Indiana Railroad. The search for God and health attracted others; healing mind and body would do little good if the soul languished in moral corruption. This too the wilderness could cure. For the pious, the Grand Rapids and Indiana Railroad constructed another stop along its hay fever and fishing line in the town of Bay View, one mile north of Petoskey.

Bay View was founded soon after Samuel O. Knapp, a prosperous businessman and lay member of the Methodist Episcopal Church in Jackson, Michigan, took his wife on the newly opened Grand Rapids and Indiana line to Petoskey in 1874. Seeking to restore his wife's failing health, Knapp brought her to the bracing climate of northern Michigan, which he knew firsthand as a mining agent for the Minnesota Mining Company. His discovery of ancient pit mines on the Keweenaw Peninsula in upper Michigan in 1848 had helped turn copper prospecting into copper fever. The discovery also added intrigue to a then unsettled dispute about the history of the Ojibwa and their relationship to the prehistoric peoples and cultures of the Great Lakes basin. In 1874, Knapp made another serendipitous discovery regarding Michigan's natural resources: in Petoskey, his wife's respiratory troubles disappeared.

Knapp was also a trustee of the newly formed Michigan Camp Ground Association, an organization of sixty thousand members in search of a suitable site in northern Michigan for an annual camp meeting. The group hoped to eventually rival the success of Martha's Vineyard Campmeeting Association. There, as many as twelve thousand people vacationing on Cape Cod or staying in one of the five hundred gingerbread cottages built on the association's thirty-four acres on Martha's Vineyard came together in worship on Sundays at an open-air, wrought iron tabernacle. The potential revenue of such a gathering piqued the interest of a number of railroad companies. Although the Pere Marquette Railway, the J. L. and S. Railway, and the Michigan Central Railway all hoped to court the business of the newly formed Michigan Camp Ground Association, Knapp convinced the board members of Petoskey's advantages over other proposed locations. First and foremost was health. Association members sought a "place where worn out bodies could be recuperated by God's own medicine, pure bracing air and good water."[62] They could find no better location, Knapp argued, than the bluff overlooking Little Traverse Bay. "Known to be a *Sanitarium* for hay fever, asthma, and catarrhal affections, . . . the whole country for scores of miles in almost every direction offers facilities for recreation which are simply endless."[63] It was also within easy reach by boat or rail from Chicago, Milwaukee, and all points south and east.

In the spring of 1876, the Michigan Camp Ground Association entered into an agreement with the Grand Rapids and Indiana Railroad to establish Bay View, a summer resort community on 326 acres of waterfront property on Lake Michigan. The railroad deeded the land and extended the line from Petoskey. For its part, the Michigan Camp Ground Association agreed to invest $10,000 in buildings and improvements on the site over the next five years and to hold its annual meetings there through 1891. By the time the association's contract expired, Bay View had far exceeded all expectations. In the 1890s, when the Western Hay Fever Association began meeting in the newly constructed Evelyn Hall, Bay View contained four hundred summer cottages, a hotel, and six large public buildings, and it regularly hosted an annual camp meeting, Sunday School Congress, training school and headquarters of the Women's Christian Temperance Union, summer university, and Michigan Chautauqua Assembly. Railroad and boat operators eagerly looked forward to Bay View events, like Big Sunday, when eight thousand visitors from neighboring resort towns and associations came to partake in a day of revivalism and religious fervor. The summer lecture series of the Chautauqua Assembly and Women's Christian Temperance Union, where the audience could listen to the likes of Dr. John Harvey Kellogg (inventor of corn flakes and director of the Battle Creek Sanitarium) extol healthful living, was always popular and drew up to twenty thousand summer attendees.[64]

Bay View became a community devoted to mental, physical, and spiritual renewal. Amid its inspiring views of Lake Michigan and surrounded by forests of hemlock and birch, hay fever sufferers, even of modest means, found a welcome home. In its early years, a Bay View life membership cost $10 and was open to any person of "good moral character," which, to the Bay View board of trustees, meant any person white and Christian.[65] Membership entitled one to the lease of a building lot, which cost an additional $2–10 per year. Members could also purchase half-fare excursion rates on the railroad lines. Such benefits were attractive to hay fever sufferers, who made up a sizeable portion of the town's residents because of recommendations of the Western Hay Fever Association. The community, in turn, had a moderating effect on the often strident class-consciousness of hay feverites. With an annual membership rate of twenty-five cents, the Western Hay Fever Association was far less exclusive than the U.S. Hay Fever Association. Through their role in the Women's Christian Temperance Union, the Sunday School Congresses, and the Chautauqua Assembly, women also shaped the character of the Western Hay Fever Association. Unlike their eastern counterpart, the Western Hay Fever Association permitted women to hold executive office. One Bay View resident, Mrs. Pope, was elected to serve as president of the association in 1896.[66]

For the Midwest's social elite who would rather flaunt their wealth and disease, the preferred and final stop on the "Hay Fever Express" was the Grand Hotel on Mackinac Island. The luxury hotel was a joint enterprise of the Grand Rapids and Indiana Railroad, the Michigan Central Railroad, and the Cleveland Steamship Navigation Company. Although it had long been a summer gathering place, Mackinac Island became the "Newport of the northwest" in the 1890s (Figure 7).[67] Drawn to the abundance of fish found in the straits in summer and to its central location within the Great Lakes, the Anishnabeg people had regarded Michilimackinac as a sacred place for almost one thousand years before it became a mecca for hay fever invalids. Through successive waves of French, British, and American settlement, Mackinac Island flourished in the northern fur trade. As the headquarters of John Jacob Astor's American Fur Company in the early nineteenth century, it was the entrepôt of the Northwest. Trappers in canoes loaded with the pelts of beaver, muskrat, otter, fox, and mink, harvested in the winter, came in the summer to trade for supplies and goods. As the fur trade declined, other bountiful resources of nature sustained the island's economy. Large populations of white fish, lake trout, pickerel, and herring supported a commercial fishing industry. In the 1870s, fish processing and shipping stations on the Michigan mainland substantially cut into the Mackinac trade, but nature had "done more for the island of Mackinac and its vicinity than any other spot on the shores of the Great Lakes," one islander observed.[68]

Figure 7. Mackinac Island became the "Newport of the northwest" in the 1890s. Passenger steam boats, such as this one, offered wealthy patrons a luxurious passage en route to the Grand Hotel, the fashionable place to stay among Chicago's smart set and exclusive hay fever sufferers. Courtesy of Mackinac State Historic Parks.

When Congress created a national park on the island in 1875 for the "health, comfort, and pleasure" of the people, a new chapter in the island's consumption of the bounties of nature began.

Mackinac National Park was the second national park to be established by an act of Congress. The first was Yellowstone. If Mackinac lacked the magnitude of nature, as captured in Thomas Moran's dramatic canvases of Yellowstone, it nevertheless possessed a magic and charm that nineteenth-century tourists prized. Layers of Indian legends became folded into every limestone rock that had been worn by time into a curious geological feature, natural arches and spiring bluffs that afforded breathtaking views of Lake Huron. It was here that Henry Schoolcraft, a U.S. Indian agent at Fort Mackinac, traded in Ojibwa stories and legends that would become the basis of Henry Wadsworth Longfellow's "The Song of Hiawatha." It was here, on top of a pillar of rock, guidebooks told, that Mechenemockenungoqua, a young Ojibwa girl, met the warrior Geniwegwon and leapt to her death when her lover, killed in battle, beckoned her from the spirit world to join him. And it was here, in "one of the purest, driest, clearest, and most healthful atmospheres," that Hygeia, the goddess of health, had placed "her temple." Mackinac Island's rich history and Indian traditions, along with its "health-restoring and health-preserving" climate and geology, were the reasons that compelled citizens and government officials to set aside half the island as a national park. Here visitors found nature, romantic and gentle in its moods and ideally suited to soothe the ills and please the tastes of high society.[69]

Tourists flocked to Mackinac Island in increasing numbers during the 1880s. The former military fort, trading post, and fishing village was ill equipped to meet the demands of the moneyed upper class strolling through its shaded glens or climbing its rock precipices. In 1886, the Grand Rapids and Indiana Railroad, the Michigan Central Railroad, and the Cleveland Steamship Navigation Company formed the Mackinac Island Hotel Company. Their plan was to build a hotel on Mackinac in the grand style that had made Saratoga Springs, New York, and Newport, Rhode Island, the resort playgrounds of the East Coast's fashionable establishment.[70] Opened at a cost of $250,000 in July 1887, the Grand Hotel quickly became the place to be among Chicago's "smart set."[71] Mrs. Berthe Potter Palmer, one of Chicago's most prominent socialites, made this "emerald gem" a favored summer playground. The society columns of Chicago and Detroit newspapers listed the whereabouts of the island's wealthy guests, while manager James Reddington Hayes, affectionately known as "the Comet," saw to it that a host of entertainments and outdoor activities (including tennis, cycling, gambling, dancing, sailing, and leisurely hikes) kept his patrons occupied. Hayes also relied upon the business of wealthy hay feverites to keep his hotel open two months beyond the summer season and to keep the Mackinac Island Hotel Company solvent. "Long after the butterflies of fashion have flown south," noted a reporter for *The 400,* Chicago's trend-setting travel magazine, "the hay fever victim finds the island a literal paradise." Hayes offered his hay fever clientele reduced rates after 1 September, and his hospitality earned him a reputation as their "patron saint."[72]

Along the northwestern shore of Lake Michigan, hay fever refuges sprang up along the tracks of the Grand Rapids and Indiana Railroad line. The various contours of class and place gave each safe haven a unique niche and appeal in the hay fever tourist trade. Yet all such places shared a common history and nature myth that evoked a more pristine time and place. Almost every railroad brochure, steamship booklet, tourist guide, or town history narrated a story that took the traveler on a journey deep into the mythic past: noble, romantic, and pure. "From the head of what is now called Grand Traverse Bay, northward," the Grand Rapids and Indiana Railroad informed its passengers, "the entire region was one vast camping-ground for Indian resorters, as tradition tells us, long before the pilgrims had landed on Plymouth Rock or John Smith had flirted with Powhattan's dusky maidens, or the Spaniards had effected the permanent settlement of St. Augustine." To walk "where every hillock is an Indian's grave," to rest under the boughs of "great hemlocks, reaching out their dusky arm, preaching immortality," or "to fish in crystal brooks . . . full of shining sparkling trout" was a nostalgic retreat that modern life had all but vanquished.[73] Hay feverites sought to cleanse themselves of civilization in the pure waters, cool breezes, and remaining forests along the Great Lakes shores and to leave behind what was, for them, the source of their disease.

Like his father before him, Ernest Hemingway looked to the streams, forests, and lakeshores of upper Michigan as a place of healing. He had come to know this land as a child, having been brought to the region because of his father's hay fever. It was a period in which the land withstood shifting human exploitation; from a place of harvest it had become a place of refuge and regeneration. In "A Big Two-Hearted River," published in 1925, shortly after Hemingway had returned from the horrors of the Great War, the protagonist, Nick Adams, steps off the train in Seney, a former logging town seventy miles northwest of Mackinac Island, only to find it gone, burned to the ground. But the river Nick had known had endured, and islands of dark pine forest remained in the burned-over land. Nick had come to escape—from what predicament of modernity, we do not know. But as he turned away from the railway and headed into the pine plains—a symbolic turn away from culture and into nature—Nick "felt he had left everything behind, the need for thinking, the need to write, other needs. It was all back of him."[74] Sleeping under the canopy of pines on the brown forest floor, cooking over a campfire, and fishing for trout in the ice-cold waters of a rushing stream, Nick settles into the rhythms of nature. Renewed feelings and sensations come as the strains of mental life and the din of civilization recede into the distant spaces of Nick's mind. Nick's relationship to the natural world echoed that of the senior Hemingway on Walloon Lake. In the imagined purity of nature and northern Michigan's distant past, Nick, the Hemingways, and many other urbanites from Midwestern cities sought respite from the anxieties of the modern world.

If place shaped illness, illness also shaped place. Seasonal residence made hay fever sufferers, like tourists in general, outsiders to the local community. But their wealth, patronage, and illness combined to make them a powerful force in town development and land use. The presence of hay fever sufferers induced hotel proprietors to keep their doors open a month beyond the traditional August end of the annual summer holiday. Hay feverites were the primary reason the tourist season in the White Mountains and in northern Michigan expanded into the early fall. Hay fever sufferers shaped town life in other ways as well, from the interior decor of hotels to the landscaping of sidewalks and streets to natural resource development. They may have been strangers to the community, but this did not stop them from becoming involved in local and regional environmental affairs.[75]

Illness made hay fever patients keenly interested in the environment around them. The combination of temperature, moisture, atmospheric conditions, and vegetation that contributed to a region's therapeutic properties was a puzzle to physicians, scientists, and sufferers. But lay knowledge, natural history, and medical geography offered clues. In the 1870s the acclaimed discoveries of Louis Pasteur and Robert Koch, who found bacteria to be the cause of deadly diseases such as anthrax and tuberculosis, led some physicians to postulate that hay fever also had a microbial origin. Such speculations, however, did not accord well with patient experience. In their view, hay fever was caused not by a specific organism but by an imbalance in the relationship between constitution and environment. To the question posed by one sufferer, "Was the disease in an incipient form in me, only waiting for an exciting cause which was found in the dust, or was it in the atmosphere?," Morrill Wyman offered an answer. An individual predisposition, found more prominently among people with indoor occupations, resulted in a susceptibility to exciting causes that acted upon the nervous system during particular seasons of the year.[76] The exciting causes that triggered hay fever were well known to hay feverites, but significant individual variation required each patient to become "his own physician." The dust and smoke associated with railroads were the bane of many a sufferer, a cruel irony given the necessity to travel by rail to find relief (Figure 8). Strong sunlight, various fruits but particularly peaches, and the fragrances of flowers could also trigger an attack during the sneezing season.[77]

Preserving the purity of place took on urgency as hay fever tourists saw and felt the beneficial properties of an exempt region wane. Contrary to the claims of town boosters and hotel owners, not every year was free from symptoms in hay fever resort areas. Every exempt place, some more than others, had years to which sufferers referred as "off color." Even Bethlehem reported such years in 1880 and 1881.[78] Two seasons of unusually hot and dry weather, coupled with the prevalence of southwest winds and smoke from forest fires that extended as far west as the Great Lakes, had left hay fever sufferers in Bethlehem, Petoskey, and Mackinac Island struggling to breathe. Considerable discussion ensued at the U.S. Hay Fever Association meeting in 1881 as to whether an asterisk should be placed alongside Bethlehem and Mackinac Island in the association's list of exempt places; such a mark indicated that an area offered only partial relief.[79]

While nature had not cooperated in these years, hay feverites were also quick to blame town councils, businesses, and citizens for a failure to maintain their places of sanctuary. Bethlehem's dusty streets became a particular point of contention during these off-color years, sparking discussions between the town's seasonal and permanent residents. Spraying the town streets with water was a common technique in the nineteenth century to damp down the clouds of dust raised by horses and carriage traffic. Resistance on the part of hotel proprietors to pay for the operation of the water-sprinkling cart, coupled with an insufficient water supply, had combined with the forces of nature to create a flash point around which tempers flared. At the town association meeting in 1880, Colonel Muckle, future president of the U.S. Hay Fever Association, accused Bethlehem citizens of being "derelict in providing for the summer comfort of boarders." Although the unsafe conditions of the plank walk and poor lighting were noteworthy neglects, "worst of all . . . was the dust," an evil upon which Muckle laid particular stress. Muckle (and

Figure 8. The dust and smoke of the railway were the bane of many a hay fever sufferer, including Mr. Weeps, the hooded gentleman on the train. Courtesy of Ebling Library, Rare Books and Special Collections, University of Wisconsin–Madison.

others) also complained of the grass and weeds left to grow in the town. To facilitate his travel between the Centennial and Sinclair Houses, he employed a man at his own expense to cut down the weeds on the south side of the road so he could walk "without irritation from the dust."[80]

Muckle's financial means enabled him to control his surroundings in ways that were not possible for everyone; other hay fever sufferers also remarked how the choice of daily paths was determined by considerations of dust, prevailing winds, and shade. To add weight to Muckle's individual voice and money, a committee was formed to pressure hotel proprietors to sprinkle the streets and to cut down weeds along the sidewalks and in the cemetery. In Petoskey too members of the Western Hay Fever Association formed an agitation committee to pressure the town government to "look after the sprinkling of the streets, cutting weeds, and other things needed for the comfort of the sufferers."[81]

Summer residents noted a marked improvement in Bethlehem's therapeutic qualities when the town finally began watering the streets. But in drought years, a limited water supply and the aggravating effects of heat and dust kindled outrage among hay fever sufferers. Only after

1894, when the precinct passed a resolution for the Crystal Springs Water Company to keep Bethlehem's reservoir full enough to supply sufficient water for fire protection and street sprinkling, did tempers cool.[82]

As the dust settled, other irritants increasingly became the focal point of concern to seasonal residents. Railroads brought hay fever sufferers to the natural sanctuaries, but they also brought unintended environmental changes. In Bethlehem, the building of a branch line of the White Mountains Railroad into Bethlehem station and an additional narrow gauge line added in 1881 to run to the town center brought a most unwelcome guest: ragweed. Able to flourish in disturbed soils, ragweed appeared along the railroad tracks just north of Bethlehem near the village of Littleton. Its presence was cause for alarm.

Ragweed, *Ambrosia artemisiifolia,* also known to nineteenth-century Americans as Roman wormwood, became a likely suspect in the growing list of exciting causes that could unleash an attack of autumnal catarrh. Natural history and botany were popular leisure pursuits among the prosperous classes in nineteenth-century America. A knowledge of plant species, as well as their times of flower, was common. Thus it was not just a predisposition of mental culture but also one of botanical knowledge that led more than one sufferer to report on the irritations provoked when he or she was walking along a road where ragweed was present. It flowered in the middle of August, grew abundantly along the seashore and in catarrhal regions, and was rarely present in the mountains. Determined to prove the plant's malicious properties, Morrill Wyman gathered specimens of the flowering weed at his Cambridge home, sealed them in a parcel, and carried them to Glen House in the White Mountains during the 1870 hay fever season. On 23 September he and his son opened the package and snorted its contents. Promptly they started sneezing, and their eyes, nose, and throat began to itch. When eight hay feverites at the nearby Waumbec House received a present of flowering ragweed from Wyman, they launched into a chorus of sneezing and wheezing fits.[83] Eight other hay fever guests, who did not inhale the pollen, looked on in amusement. When ragweed eventually arrived in the White Mountains on its own accord, the community mobilized an effort to exterminate the "baneful weed" but without success. In 1886, Dr. Morrell Mackenzie observed, "The spread of this pest is simply marvelous."[84]

By the turn of the century, ragweed had become symbolic of all that hay fever tourists sought to escape. Like hay fever, it came to be regarded as a product not of nature but of civilization. Other pollen-bearing plants that followed the plow also came under attack. "Everything around the patient [was] saturated with . . . poisonous emanations [of] corn, peas, fodder, and other farm products," wrote James Bell in the U.S. Hay Fever Association's prize essay of 1887.[85] "Increasing cultivation and growing use in parlors and dining rooms of fragrant flowers" were also, in the opinion of the Reverend John Peacock, "vitiating the natural purity and exemptiveness of the atmosphere of Bethlehem and its vicinity."[86] The proliferation of vegetable and flower gardens in the early 1890s prompted the U.S. Hay Fever Association to pass a resolution urging Bethlehem citizens to restrict "the planting of corn and other pollen-bearing vegetables" on the north side of the town at some distance from the street (Figure 9).[87] "Improvements and other civilizing changes" in Bethlehem were, in the opinion of some, "diminishing its immunity from Hay Fever."[88] "Even in the very best of resorts," observed the naturalist and theologian Samuel Lockwood, "unless Nature has been left to her virgin forms and moods," complete relief could no longer be found. The proximity of local industries, "whether of agriculture or other pursuit," had affected Bethlehem's exemptive qualities.[89]

In contrast to the situation in tuberculosis sanatoria, where an emphasis on nutritional treatments proved beneficial to the surrounding farm communities, hay feverites in the White Mountains found little to praise in the region's agricultural land use.[90] Colonel John Ward of Louisville, Kentucky, believed that the relative absence of cultivated farmland in northern Michigan, as compared to the White Mountains, made it a preferred hay fever haven.[91] The disdain expressed toward agriculture and the value placed on "virgin Nature" by members of the U.S. Hay Fever Association were deeply embedded in the artistic and literary production of the White Mountains as a place to experience the natural sublime. In July 1858, en route from Concord, Massachusetts, Henry David Thoreau reflected on the view northward as he approached the White Mountains, where Glen House served as his base camp: "A dozen miles off seemed the boundary of cultivation. . . . I felt near the edge of a wild and unsettlable [*sic*] mountain region." Thoreau's journal entry alludes to an increasing divide between "civilization" and "nature," between the city and the country; his writings helped to establish the divide later in the nineteenth century.[92] In grounding the pastoral ideal in leisure, in natural history pursuits such as botany and bird watching,

Figure 9. Sinclair House, 1870s. The presence of vegetable gardens, seen here, became of increasing concern to Bethlehem's seasonal hay fever residents. Courtesy of Bethlehem, New Hampshire, and Bondcliff Books.

Thoreau denigrated agriculture as the most desirable relationship with nature. Later nature writers, like John Burroughs, looked to Thoreau and to his mentor, Ralph Waldo Emerson, in crafting an Arcadian landscape. It was a place of wilderness where the educated classes might find solace and regeneration from the throngs of city life. If pristine nature was the antidote to civilization by century's end, then surely it would benefit those suffering from the malady for which modern urban life was held responsible: hay fever.[93]

Agriculture was not the only White Mountain industry that hay fever tourists regarded with suspicion. Commercial logging also became of increasing concern to hay feverites in the last two decades of the nineteenth century. Large pulp and paper producers had begun harvesting second-growth conifers at a devastating rate (Figure 10). The dense forests and high mountains of the region created a barrier to southerly winds, which always "caused trouble" whenever they appeared in an exempt district. Many hay fever sufferers believed that "the passage of the south winds through [the] forests robbed [them] of their noxious elements." "Forests," Edmund Hoyt, secretary of the U.S. Hay Fever Association, told its members, "bore an important relation to the disease and should not be destroyed."[94] This sentiment accorded well with popular medical opinion of the 1880s, which ascribed therapeutic value to coniferous forests, particularly in the treatment of malaria and consumption.[95] But it also reinforced the growing use of nature through leisure by America's middle and upper classes, to which hay feverites largely belonged. "Trees and forests," one sufferer noted, "are worth more, financially, than any garden vegetables or fragrant flowers that can be grown in their place. The preservation and even cultivation of trees would do much to retard the deterioration of this region, from a Hay Fever point of view, without lessening the pleasure of those who come for other reasons."[96] Consuming nature for health and pleasure went hand in hand.

Fearing that "deforestation, cultivation, and civilization" would soon "drive Hay Fever people to Maine, the British Dominion, and Europe as their only resort," the U.S. Hay Fever Association called upon its members to become active in a campaign launched in the 1890s by the Appalachian Mountain Club (AMC), the Society for the Preservation of New Hampshire Forests (SPNHF), and the New Hampshire Forestry Commission to establish a federal forest reserve in the White Mountains.[97] Attributing the region's health-giving qualities to the "purity of its atmosphere," U.S. Hay Fever Association officials pointed to the benefits of New Hampshire forests and the economic loss their wanton destruction would have on the region's tourist industry.[98] Only through the active cooperation of "Hay Fever sufferers in the Forestry movement" could the exemptive qualities and identity of the region be assured.[99]

The shared interest in forest preservation among members of the U.S. Hay Fever Association, the AMC, and the SPNHF was grounded in more than just a common political cause. It was also rooted in the material and social relations that defined the White Mountains as a hay fever refuge. Amid the descriptions of hikes and new proposed trails, health—

GLEN HOUSE, WHITE MOUNTAINS, N. H.

The largest and most desirably located House in the region, and the only one where from every window of its four hundred feet front, and its broad piazza (extending to double its former width), the highest mountains in New Hampshire are distinctly seen from base to summit, viz., Mts. Washington, Jefferson, Adams, Madison, &c. A sure relief is here obtained for catarrhal complaints, hay fever and rose cold. A large Farm from which vegetables are obtained, and a fine Dairy. Its Cusine will compare favorably with the best Hotels in the United States. The "Glen" is reached from Gorham, N. H., on the Grand Trunk Road by eight miles staging; from the Glen Station on the Portland and Ogdensburg Road fifteen miles; from the Mount Washington R. R., eight miles.

Figure 10. Advertisement and photograph of Glen House, 1879. Many hay fever sufferers feared that deforestation, evident in the photograph, was destroying the qualities of Bethlehem as a hay fever refuge. *White Mountain Echo,* 9 August 1879, 11, and Moses Sweetser, *Views in the White Mountains* (Portland, Maine: Chisholm Bros., 1879).

and hay fever in particular—occupied a persistent theme in the decade-long correspondence of AMC members Edith Cook, Lucia and Marian Pychowska, and Isabella Stone. A hay fever sufferer and member of the AMC for forty-one years, Isabella Stone first began traveling to the White Mountains in 1863. According to nineteenth-century medical accounts, Stone was, as a woman, among the minority of hay fever sufferers. But just as class and race barriers at hay fever resorts created the impression that hay fever was a disease of the white middle and upper classes, regulations of the U.S. Hay Fever Association reinforced the notion that hay fever was an illness that afflicted males in greater numbers. The U.S. Hay Fever Association restricted the participation of women to the advisory board; only males were entitled to full membership on the executive committee. In the mountains, however, Isabella found relief from her disease and freedom from the limiting gender roles of the U.S. Hay Fever Association and New England society. Whether ascending Mount Lafayette, clearing the first trail to Bridal Veil Falls in Franconia Notch, or guiding a party up to Loon Pond Mountain, Stone took delight and pride in adventures that were not open to her at her home in Framingham, Massachusetts. The adventures were made possible only through her seasonal affliction. And they were all the more reason to ensure that the forests and mountains in which she and others found freedom and relief be preserved.[100]

To the region's summer residents, the argument for preserving New Hampshire's forests in the name of health was equal to standard conservation arguments for clear streams, stable soils, and sustainable timber. What's more, such arguments were not made only by the region's health seekers. In its 1891 report to the New Hampshire legislature, the Forestry Commission listed the "life-giving and health restoring qualities" of the White Mountain region as one of four primary reasons to protect and preserve the state's forests.[101] SPNHF lecturer John D. Quackenbos similarly praised the therapeutic benefits of "the rank scenting ozones and balsamic aromas" of New Hampshire's evergreens to help enlist support for public ownership of the White Mountain forests.[102]

Naturalists and hay feverites such as Samuel Lockwood well understood that if the White Mountains were to remain a favored refuge, more than individual testimonies to the power of place were needed. To attract hay fever exiles—particularly as dissenting voices grew alongside Bethlehem's weeds—would require a bigger effort. Other places, including Petoskey, Mackinac Island, Denver, and Colorado Springs, vied for the hay fever tourist trade. In 1896, Denver's Chamber of Commerce and Board of Trade invited the U.S. Hay Fever Association to move its annual meeting to the "Queen City of the Plains." After all, Denver was known for its curative power over asthma and consumption. It was a reputation backed not only by individual testimonies but also by scientific studies on Colorado's climate and health that had been conducted by Charles Denison, president of the American Climatological Association.[103] In 1888 Lockwood himself had turned the efforts of the U.S. Hay Fever Association toward what he hoped would become a comprehensive scientific study and comparison of the atmosphere in exempt and nonexempt regions. Enlisting the support of laymen, Lockwood endeavored to gather meteorological records of temperature, wind velocity and direction, and humidity and barometric changes. In addition, he hoped to include a microscopic analysis of atmospheric particles, along with the experiences of patients under their local influence, to arrive at trustworthy results "on the line of comparative pathology." Scientific proof of the purity of Bethlehem's atmosphere would be "highly beneficial to the place," the *White Mountain Echo* observed as it urged hotel proprietors to contribute to the association's scientific fund.[104]

During three consecutive seasons, Lockwood conducted a comparative microscopic analysis of the air around three White Mountain resorts (the Maplewood, the Twin House, and the Waumbec) and his nonexempt home in Freehold, New Jersey. Daily catches from 16 August through 20 September were collected on slides coated with glycerin, and their contents were compared. Of the fifty slides Lockwood collected at the Maplewood in the first season, only two showed any signs of pollen. The bulk of the collected material was minute particles of wood, mineral dust, and the occasional scales of butterflies and moths. In contrast, the slides from Lockwood's home held hundreds of ragweed pollen grains, in addition to greater amounts of road dust. The slide collections showed that the amount of pollen tapered off in late September. Compared to Freehold, the average daily temperature in the White Mountains was approximately ten degrees cooler, and the air was decidedly less humid. Although Lockwood was unable to undertake a chemical analysis of Bethlehem's air, he believed that the cool, dry air, in addition to the "terebinthine effect from the balsams which clothe the mountains" made the air "markedly tonic." This tonic air, coupled with the smaller quantity of vegetable matter and comparative absence of pollen, went far, Lockwood argued, in explaining the hygienic qualities of the White Mountains as a valuable hay fever resort.[105]

When the Honorable J. B. Walker, president of the Forestry Commission of New Hampshire, addressed the future of the White Mountains in 1892 during the initial campaign for the creation of a national forest reserve, he found an enthusiastic audience among the region's hay fever residents. The $5 million in estimated income from summer tourists visiting the region exceeded the annual income derived from New Hampshire farmers. To Walker, the region's economic future rested not in timber and farms but in the preservation of its forests and natural scenery.[106] His vision of land use was heartily endorsed by hay fever exiles and confirmed by Lockwood's study. Only through the preservation and cultivation of New Hampshire's forests and the establishment of a state park could the region remain an exempt area.

Elsewhere, changes in the land seemed to be despoiling the natural places of refuge to which hay fever tourists flocked. In the opinion of the Reverend John B. Sewell, "reckless axe and the denuding fire" had left Petoskey a less desirable destination for hay feverites at the close of the nineteenth century.[107] In "The Last Good Country," Ernest Hemingway disparaged the relentless harvesting of northern Michigan's conifer forests; it had taken place in earnest a decade before he was born and had jeopardized the region's therapeutic benefits for hay feverites like his father. The fictional Nick Adams and his sister Littless climb through a slashed area of hemlocks, among piles of brush and rotting gray logs left by the barkpeelers of businesses like Petoskey's Michigan Tanning and Extract Company. Irritated by the heat and the "pollen from the ragweed and the fireweed," Littless is overcome with sneezes. "Damn slashings," she says to Nick. "I hate them. . . . And the damn weeds are like flowers in a tree cemetery if nobody took care of it."[108]

Renewal, of both tired lands and exhausted people, was the motivation for the establishment of Michigan's permanent forestry commission in 1899. It had also been an impetus in the preservation of New Hampshire's forests. Michigan's first forestry commission president, Charles Garfield, saw a great future for northern Michigan as a new century dawned. "Great areas of trees all up and down this beautiful state, protecting head waters of our rivers, making use of our unfertile sands, giving variety and beauty to our gentle hills and refreshing the weary, whether human or otherwise, with nature's quiet cathedrals," would come to be, Garfield assured citizens. The state would renew "its waste places with forest life." The path to the future turned back toward a more pristine time and place. Hay feverites at the turn of the century sought to reenact the pilgrimage to a natural tabernacle to heal body and soul. They hoped to purify the landscape and themselves from the malady of civilization. Their enemy was both without and within.[109]

By the mid-twentieth century, the automobile had transformed hay fever holidays into an experience of mass consumption. In the seventy-five years since hay fever resorts made visible an affliction that was once thought to be confined to prosperous white, largely male professionals who lived in urban centers east of the Mississippi, the disease had moved across the boundaries of geography, class, and gender. The spread of hay fever across the North American landscape coincided with changing causal explanations, treatments, cultural anxieties, and environmental conditions. Nevertheless, the places that wealthy hay feverites had initially helped to build continued to thrive on economies catering to health and recreation long after the grand hotels and conspicuous consumption of the Gilded Age were but a distant memory.

Hay fever tourists had largely succeeded in helping preserve the White Mountains as a hay fever refuge. In July 1951, the *Saturday Review of Literature* featured a checklist of hay fever resorts. These were places where an estimated 4 million hay fever sufferers might find both recreation and relief. The White Mountains occupied a prominent position in the list of East Coast hay fever getaways. Just four years before the magazine feature, New Hampshire's Department of Health had conducted a statewide pollen survey that demonstrated the absence or relatively light presence of ragweed and other hay fever pollens in the White Mountain region. Franconia Notch State Park and the surrounding 780,000 acres of the White Mountain National Forest acted as a barrier against the ragweed and grass pollens carried by the prevailing northwest winds from Canada, although southwest winds sweeping up through the farmlands of the Connecticut Valley could still make for an occasional off-color year.[110]

The Straits of Mackinac and the fashionable beach colonies along Lake Michigan's Little Traverse Bay also remained among the most popular hay fever vacation destinations after World War II. Ca-Choo Clubs and Un-Ca-Choo Clubs proliferated along Lake Michigan's northern shore, continuing the tradition of friendship, recreation, and entertainment in Michigan's "veritable summer playground" that had begun almost seventy years before.[111] In the intervening years, forestry, conservation, and game management efforts had restored the cutover lands of northern Michigan. It was once again the kind of sportsmen's paradise and recreational wonderland that had first attracted wealthy hay feverites to the area after the Civil War. By 1949, with 3.6 million acres in state forest and recreational lands, Michigan had become a national leader in resource conservation and management.[112]

Business Week predicted that a record number of "refugees from the hay fever belts" would make their way to popular summer resort areas in the period after World War II.[113] As had a previous generation, this postwar cadre of hay fever tourists enjoyed newfound prosperity and leisure time that made their disease an almost welcome seasonal event. Attracted to the healthful and recreational benefits of nature, they flocked to places like Mackinac, where "centuries ago the Indians regarded the 'Magic Island' not only as the favored spot of the Great Spirit but as a sanctuary where illness and disease fought a losing battle." It was also a place where nature still served as "the best ministrator [*sic*] to those distressed by the effects of unwelcome pollen."[114] Antihistamines and other biomedical wonders would soon offer a technological fix to an illness that modern civilization had begun. Until then, however, the preferred remedy was an annual retreat to nature. A hay fever holiday might be less convenient, but to those who could afford it, it was certainly more fun.

CHAPTER

2

When Pollen Became Poison

Ragweed was here thousands of years before the *Mayflower* touched the shores of the New World, and I predict that it will be here as many thousands of years after the human race has ceased to trouble this planet.

—Roger Wodehouse, 1940

Immunology, the scientific study of the body's immune system, was a fledgling science in the early 1900s, but it had already yielded great promise in the prevention and cure of some of the most deadly epidemic diseases threatening public health. In the closing decade of the nineteenth century, vaccines and antitoxins against anthrax, rabies, diphtheria, and tetanus came out one after another. Leading research laboratories around the world, including the Pasteur Institute in Paris and Robert Koch's Institute for Infectious Diseases in Berlin, were the sources of these great advances. In the United States, drug companies and laboratories, as yet unregulated, swiftly mass-produced these new biological products, often with little regard to possible and already known dangers associated with their use. In these heady days of biomedical triumphs and breakthroughs, it looked as though human diseases would soon be eradicated.

The widespread adoption of vaccination and serum therapy in the opening decade of the twentieth century presented physicians with many opportunities to witness a new host of immunological reactions firsthand. In their Vienna practice, for example, pediatricians Clemens von Pirquet and Béla Schick observed numerous cases of serum sickness after injecting young patients with horse antitoxins, which in 1905 were cutting-edge therapy in the treatment of diphtheria and tetanus. Symptoms such as high fever, achy and swollen joints, and itchy skin rashes were not uncommon. In some cases, the reactions were severe. Patients had been known to die suddenly after being given diphtheria antitoxin prepared from horse serum. The organism, it seemed, could respond to foreign substances in ways harmful rather than beneficial to itself (to put it mildly).[1]

Other phenomena, observed in the laboratory, displayed similarities to the side effects of serum therapy witnessed by doctors in the clinic. In 1902, the French physiologist and future Nobel laureate Charles Richet, along with his colleague Paul Portier, accidentally produced an anaphylactic reaction in dogs while attempting to immunize the animals against marine toxins. The two physiologists had become intrigued by the poisonous properties of certain marine organisms while sailing in the Indian Ocean on board the *Hirondelle II,* the yacht of Prince Albert I of Monaco. Foremost among these was the Portuguese man-of-war, a jellyfish common to tropical and subtropical waters. The paralyzing sting of this hydroid can immobilize small fish and inflict a painful skin reaction in the unsuspecting swimmer. When Richet and Portier injected dogs with its poison, the animals showed no signs of distress. If, however, they administered a second injection three to four weeks later, the dogs underwent a violent reaction and died within thirty minutes. Richet had not built up immunity in dogs but had instead created a condition of acute sensitivity; he labeled this condition anaphylaxis.[2]

The immune system played a role in more than protection against disease. Serum sickness and anaphylaxis highlighted its role as a source of illness. In 1906, von Pirquet and Schlick coined the term "allergy" to describe conditions where an immune response of the organism inflicted harm rather than benefit on the organism. Derived from the Greek words *allos,* meaning other, and *energia,* understood as energy or activity, allergy referred to the "altered reactivity" of the body to a foreign substance, be it bacteria, protein, pollen grain, or poison.[3]

In the body's response to introduced foreign substances, both serum sickness and anaphylaxis presented parallels to the conditions of hay fever, asthma, and certain skin diseases, such as urticaria, that had been observed by clinicians. The German physician Alfred Wolff-Eisner suggested that pollen, like the poison of the Portuguese man-of-war, could trigger an immune response that prompted a violent immunological reaction in certain individuals. By putting drops of a pollen solution into the eyes of his asthma and hay fever patients, Wolff-Eisner was able to produce symptoms commonly experienced by sufferers during the hay fever season: red, swollen, and itchy eyes. His findings, published in 1906, and his explicit association of hay fever with anaphylaxis, marked an important shift in the medical understanding of hay fever. Previously seen as a disease of the nervous system, hay fever came to occupy a place in the new biomedical frontier of the body: the mysterious, little-explored immunological landscape.[4]

Asthma shared a place alongside hay fever in this new medical geography of body and illness. In 1910, Samuel Meltzer, chief of the department of physiology and pharmacology at New York City's Rockefeller Institute, a powerhouse of American biomedical research, noted that laboratory guinea pigs in a state of artificially induced anaphylactic shock behaved similarly to humans suffering an asthma attack. In both instances, constriction of the bronchial airways made it difficult, if not impossible, for the animal or patient to breathe. Both asthma and hay fever, Meltzer concluded, were anaphylactic responses triggered by hypersensitivity to a protein substance invading the body. By the second decade of the twentieth century, researchers and physicians had reconstituted hay fever and asthma into allied allergic diseases.[5]

With the shift in medical understanding of hay fever as a disease of the immune rather than the nervous system, treatment also shifted and expanded from the therapeutic wilderness to urban clinics. Modern medicine made it possible, through the production of pollen vaccines, to "remain in town and at work" during the spring, summer, or fall hay fever seasons.[6] But the promise of relief did not emerge solely from the immunological laboratory. It also came from a partnership between doctors and botanists created by the necessity to address the ecology of hay fever as a disease. To precisely understand the relationship between the prevalence of hay fever and the presence of certain plants, botanists and doctors mapped the abundance and distribution of grasses, trees, and other plants in the city. They sampled and collected pollens for diagnostic and therapeutic use in allergy clinics; the knowledge gained from these analyses, in turn, aided in the study of the ecology, evolution, and changing distribution of North American plants over time. Through the spaces of the laboratory, clinic, and field, new ecological relationships between people and plants were forged.

In the urban setting of the clinic, where the definition, study, and treatment of allergic disease came into being, wild nature occupied a different place in the ecological order than it had in hay fever resorts. To the hay fever tourist, immunity was found, not in a bottled vaccine, but in the little understood yet confirmed experience of healing and relief in nature. In the city, nature's wildness was a threat to urban public health. Domesticated in city parks, planted to create tree-lined streets, and carefully manicured in lawns and gardens, nature was allowed to thrive. This was nature's proper place in the ordered spaces of the city—managed and controlled. But elements of uncontrolled nature took root in the cracked pavements, vacant lots, and wastelands of the modern metropolis. These were the places where hay fever plants like ragweed flourished. To the city's better-off citizens, these neglected neighborhoods were unsightly, unhealthful, and often unlawful places in need of eradication or reform. Because it thrived on the urban fringes, ragweed, a North American native plant, became associated in the minds of doctors, sanitary engineers, and public health reformers with other urban "weeds"—the tramps, transients, and urban poor who eked out an existence on the margins of civilization. Ragweed's identity became marked by its ecological niche: a "slum dweller, preferring to live in . . . city dumps," a "river rat," a "squatter on vacant property."[7] Out of place in the urban environment, wild nature became the enemy in the initial combat of American cities against allergic disease.

It all began with pollen. In Germany in 1904, Dr. William Dunbar marketed the first antitoxic serum for the treatment of hay fever. The émigré doctor, born in St. Paul, Minnesota, found little relief from his allergies when he moved to Germany to take up a post as director of the State Hygienic Institute in Hamburg. Numerous observations and experiments led Dunbar to believe that hay fever was caused by the body's reaction to a toxic protein produced by the pollen of certain plants. To determine what species were toxic required great detective skills. In an effort to determine the abundance and kinds of pollen in the air around Hamburg, Dunbar and his colleagues collected samples by placing microscope slides on the institute roof and in the botanical garden. They also tested the pollen collected from 125 different species of plants for reaction on both hay fever sufferers and normal subjects. And they tested themselves. In one human trial, Dunbar and a student assistant gave themselves a hypodermic injection of an isolated pollen toxin diluted in water. Dunbar nearly asphyxiated himself. His student broke into a profuse sweat, his face swelled, his ears turned bluish red, his pulse raced, his breathing became rapid and labored, and skin eruptions broke out over his entire body.

Terrified by the results, the two swore off future experiments that involved active immunization, the process of building up immunity by exposure to gradually increasing quantities of allergen. Instead, Dunbar developed a technique based on passive immunization, which had led to the successful prevention of diphtheria. By injecting young thoroughbred horses with pollen toxin in increasingly large doses, Dunbar produced what he believed to be an antitoxin in the horse's blood that neutralized the pollen's allergenic effects. Marketed by Schimmel and Company in Germany, Dunbar's horse serum—Pollantin—was available in liquid, powder, or salve form. Applied topically to the eyes or the nasal membranes of a hay fever patient, Pollantin offered complete relief to over 50 percent of hay fever patients who used the product in advance of the hay fever season. But its use in Dunbar's native land was of limited value. In the United States, most patients succumbed to fall hay fever—autumnal catarrh—a form of the illness absent in Europe. And many of the offending plant species in North America had no counterparts in Europe. To capture the American market, Dunbar prepared a separate Pollantin extract sold by Fritzsche Brothers, a New York chemical manufacturer specializing in essential oils, perfumes, and drugs. This new extract used German horses and American plants to produce an antitoxin serum that the U.S. Hay Fever Association found to be only partially effective against the offending pollen of ragweed, the first and most notorious hay fever plant to be identified with the fall hay fever season.[8]

The limited success of passive immunization, in addition to reports of patients going into anaphylactic shock after using Pollantin, led other investigators to follow the path of active immunization, from which Dunbar had recoiled. The first published account of pollen desensitization through direct injections of pollen extract appeared in 1911 in the British medical journal *Lancet.* The authors were two physicians, John

Freeman and Leonard Noon, who had studied together at the Pasteur Institute in Paris and were now working in the Inoculation Department of St. Mary's Hospital in London. Freeman and Noon had inoculated a small number of hay fever sufferers in the fall, winter, and spring with increasing doses of pollen extract derived from timothy. This common fodder grass, found in meadows and along waysides, flowers between June and August in England. Freeman and Noon found a "distinct amelioration of symptoms" in the patients treated. Enthused by their initial success, they expanded their study to eighty-four patients. Within a decade, their work had established St. Mary's as the center for the clinical diagnosis and treatment of allergy in Great Britain.[9]

Unbeknown to Freeman and Noon, active immunization had been tried and embraced by American physicians in Boston, Chicago, New York City, and other locations both before and shortly after Freeman and Noon's first *Lancet* publication. In Chicago, Rush Medical College professor of medicine Karl Koessler started injecting hay fever patients with ragweed extract in 1910, hoping to build their immunity against the onset of fall hay fever. Work on the pathology of bronchial asthma and its treatment by bacterial vaccines had steered Koessler toward hay fever. Questions of immunity linked the two diseases. In Koessler's opinion, hay fever was not the result of a true toxin—as Dunbar maintained—but was instead the consequence of an acquired or inherited sensitivity to pollen accompanied by anaphylaxis. Koessler argued that Dunbar had misdiagnosed the symptoms when he himself had nearly suffocated and had almost killed his assistant. Dunbar had injected a pollen extract one hundred thousand times the strength Koessler used on his patients. It was a case, Koessler believed, not of acute poisoning, but of anaphylactic shock.

In immunizing his own patients, Koessler first tested the sensitivity of a patient to ragweed pollen. He began by placing a drop of an extremely dilute solution of ragweed pollen—approximately one ten-millionth of a gram—into the patient's eye. The smallest dose that still produced a bloodshot eye served as the baseline for the strength of therapeutic inoculations. These were administered at intervals of four to ten days. Of the forty-one patients Koessler treated by active immunization, thirty-three showed a marked improvement. Koessler believed that the real test of the therapy would be "a railway or an automobile journey through flowering meadows." Although confident in his treatment's effectiveness, the Chicago doctor thought only four of his patients had built up sufficient immunity to withstand such a direct enemy assault.[10]

Given immunotherapy's initial success, Koessler remarked that "it [would] not be long before the commercial manufacturers of vaccines see 'the great advantage and benefit' of this treatment." "Hay fever vaccines," Koessler predicted, "will be praised and advertised and put up so attractively that their use will become universal and soon—universally discredited." The first part of Koessler's prediction quickly came true. In less than five years after Freeman and Noon's first publication on desensitization in 1911, pharmaceutical companies such as Lederle Laboratories, Parke-Davis, and Abbott Laboratories were marketing commercial pollen extracts for the diagnosis and treatment of hay fever. Lederle's pollen vaccine alone was in use by six hundred physicians in 1915.[11]

Koessler forecast the widespread use of pollen vaccines with an accuracy that a meteorologist would envy. His prediction that rapid commercialization would undermine therapeutic credibility also proved correct. Lederle Laboratories claimed an 83 percent success rate for its vaccine as a prophylactic against hay fever, but such bold claims did not hold for long. In a letter to physicians written around 1915, Lederle described its product as a "carefully standardized vaccine from the pollen of timothy, red top, June grass, orchard grass, wheat, sorrel, dock, daisy, maize, ragweed, goldenrod, all of which are known to be important factors in causing hay fever in the spring, summer and fall. Hence our vaccine offers protection against each individual pollen mentioned, as well as against all the pollens collectively." Outside the northeastern United States, however, where Lederle Laboratories was located, the Lederle pollen vaccine proved of little use. Dr. Grant Selfridge, a San Francisco ear, nose, and throat specialist and the first physician in that city to conduct clinical research on the diagnosis and treatment of hay fever, discouraged the use of commercial vaccines because they failed to take into account the regional specificity of hay fever as a disease. In comparing the plant species of Lederle's pollen vaccine to plant species prominent in the San Francisco Bay Area, Selfridge found "no botanical relationship between them."[12] What good was it to vaccinate a patient in California against ragweed when the plant had yet to make its way west of the Mississippi? Other physicians soon reported similar cases in which patients suffering from hay fever were tested with negative results, only to learn that the extracts that had been used were made of pollens collected over "a thousand miles from the patient's home."[13]

Pharmaceutical companies like Lederle attempted to establish a universal, standardized vaccine for the United States. Their efforts were based upon their success in marketing other biologicals to combat infectious diseases. Unlike microbes, however, pollen was not readily standardized. Plant pollen varied according to region, season, and local meteorological and environmental conditions. But Lederle's pollen vaccine was developed without this understanding for several reasons. First, Lederle sought a universalized product—one remedy that worked for everyone—and so began with a misstep—that is, the assumption that a universalized product was possible. Second, the historical geography of hay fever, which situated the ecology and economy of the disease in the northeast, misled Lederle. Lederle scientists thought they knew hay fever, but it turned out they only knew hay fever in the northeast.

Morrill Wyman's widely circulated 1872 treatise, *Autumnal Catarrh,* which included maps of the geographic distribution of hay fever in the United States, conveyed the impression in the late nineteenth century that hay fever was a disease rarely found west of the 100th Meridian. But an inquiry sent in 1916 to state boards of health by the American Hay Fever Prevention Association (based in New Orleans) began to produce a different regional picture. The California State Board of Health, for example, noted that although hay fever was present in the Golden State, ragweed, the chief offending plant among hay fever sufferers east of the Mississippi, was not. Which plant pollens contributed to the disease in California remained a mystery. Through these regional comparisons physicians soon became convinced that the causal link established "between plant pollens and . . . allergic diseases, such as asthma and [hay fever]" made "careful study of regional botany [an] essential requisite for . . . intelligent treatment."[14]

In the early establishment of allergy clinics, many physicians opted to work closely with botanists familiar with local flora to develop their own pollen extracts. A less attractive alternative was to rely upon pharmaceutical companies, which gave no thought to the region where their "standardized" vaccines were to be sold.[15] One of the earliest allergy clinics to employ a botanist in the development of pollen therapy was that of Boston physician Joseph Goodale. During the 1890s, inspired by his interactions with Robert Koch at the Institute for Infectious Diseases in Berlin, Goodale worked on the preparation of diphtheria antitoxin for the Massachusetts State Board of Health. Reports of asthmatic patients who died from anaphylactic shock after being injected with diphtheria antitoxin led Goodale to develop a test to assess the danger of antitoxin derived from horse serum to patients with asthma or hay fever. To determine a patient's hypersensitivity to the horse serum, Goodale made a superficial scratch on the patient's earlobe, where one drop of diphtheria antitoxin would be rubbed in, and also placed one drop in the patient's nostril. In patients with asthma triggered by contact with horses, a localized swelling of the skin and mucous membrane developed, accompanied by symptoms associated with hay fever. Patients not allergic to horses and with bronchial asthma and hay fever showed no reaction. Such trials prompted Goodale to recommend the use of scratch tests on patients with a history of "horse asthma" before inoculation with the diphtheria vaccine.[16]

In 1914, as an assistant physician at the Throat Clinic of the Massachusetts General Hospital, Goodale extended his studies of anaphylactic skin reactions to fifty-eight hay fever patients who had come to the clinic seeking relief from their seasonal symptoms. Goodale displayed a keen knowledge of the distribution and seasonality of New England plants in his clinical research. (His father, George Goodale, was the Fisher Professor of Natural History and Botany at Harvard and had encouraged Joseph's interest in botany from an early age.) The range of symptoms among Goodale's hay fever patients varied. Some were affected only in June, others in August, and others had symptoms that lasted from June until the first frost. Using a pointed knife, Goodale made a scratch, roughly one-eighth inch long, in the patient's arm. An alcohol extract of pollen was rubbed into the scratch with the knife blade. Goodale had selected twenty-nine species of grasses, ragweed, and other common wind-pollinated plants that flowered in the spring and summer in New England. The size of the welt that appeared offered a measure of the patient's susceptibility to different kinds of pollen. Once the list of exciting pollens was determined for each patient, Goodale began an individually prescribed treatment of active immunization. Through such diagnostic techniques, a pattern of early, middle, and late summer hay fever appeared that corresponded to the successive flowering times of New England trees, grasses, and other plants.[17]

As clinicians like Goodale tested and treated more and more patients, their need for botanical expertise, particularly in the area of plant taxonomy and biogeography, grew. Differential susceptibilities of patients to pollens, locale, and season demanded that clinicians gain a more precise understanding of the seasonal flowering, ecological habitat, and taxonomic relationships of plants. In 1915, Roger Wodehouse, a Master's student in Harvard's Laboratory of Plant Physiology, assisted in the collection and isolation of proteins from ragweed pollen for use in Goodale's expanding clinic. Upon the completion of his degree, Wodehouse went to work for Goodale as a protein chemist to improve upon the preparation of pollen extracts. With the rapid growth of allergy as a clinical field, Wodehouse found his knowledge of plant taxonomy and biochemistry in great demand. In 1917, Isaac Chandler Walker, credited with founding the first allergy clinic in the United States, recruited Wodehouse to conduct research on plant and animal proteins for a study on bronchial asthma and hypersensitivity that was under way at Boston's Peter Bent Brigham Hospital. One year later, Wodehouse left to take a research position at Arlington Chemical Laboratory in Yonkers, New York, a prominent manufacturer of hay fever and asthma drugs, while he simultaneously pursued a Ph.D. in botany at Columbia University.[18]

Medicine became a lucrative means by which Wodehouse could pursue his interests in plant taxonomy. His dissertation on the morphology and phylogeny of pollen grains helped to answer key questions with which allergists had struggled in developing treatments for their patients. For example, would a pollen extract of ragweed provide immunity to a patient sensitive to goldenrod? The two plants are different species, yet both are members of the huge composite family of flowering plants. Did extract of eastern black walnut desensitize a patient to the closely related but geographically distinct species of California black walnut that plagued hay fever sufferers on the West Coast? The answers to such questions depended upon a precise understanding of the evolutionary relationships of hay fever plants. Such understanding brought clinical, biochemical, and microscopic evidence to bear on taxonomic research. While grasses were primarily responsible for the onset of early summer allergies, plant species in the composite family—ragweed, marsh elder, cocklebur, and sagebrush, among others—were the largest offenders in the late summer allergy season. At Arlington Chemical, Wodehouse embarked on a taxonomic study of the composite family, which includes more than twenty thousand species. His effort, the first of its kind, used pollen to map evolutionary family trees (Figure 11).[19]

Such botanical knowledge was tremendously important in helping allergists predict and analyze the allergenic properties of pollen across different but related plant species. In 1916, for example, Robert Cooke and his colleague Albert Vander Veer claimed to have successfully immunized patients allergic to a variety of grasses, including orchard grass, June grass, and sweet vernal grass. They did so at their New York Hospital clinic by using an extract derived solely from timothy pollen.[20] Other doctors disputed Cooke's claim that a pollen extract from a single grass species could immunize a patient against offending species in the entire grass family.[21] But plant taxonomy supported Cooke's medical findings. Grasses, Wodehouse argued, constituted "the most compact and closely knit larger families of the flower plants." Among them, he continued, "the hay-fever excitant is strictly a family characteristic." If true, it greatly simplified therapeutic treatment of patients suffering from grass allergies.[22]

While botany assisted allergy diagnosis and treatment, allergic human bodies also furnished botanists with new methods of plant classification. Serum isolated from the blood of hay fever patients gave botanists a biochemical means to test "the phylogenetic or 'blood' relationships of plants."[23] Take, for example, a person allergic to ragweed and timothy. Serum isolated from this person is injected into the skin of a normal subject. The injection makes the localized area of normal skin temporarily sensitive to the topical application of ragweed and timothy pollen. Desensitizing the skin to ragweed does not affect the skin's allergic reaction to timothy because the antigens in ragweed and timothy pollen are different. To botanists, such differences in skin reaction gave a measure of the degree of relatedness or difference among plant species.

In some cases, allergic responses challenged botanists to revise their classification schemes. Classification of the composite family was a case in point. While modern botanists were apt to place the ragweed tribe (a group of related genera) outside the composite family based on its traits of wind pollination, such a view only misled clinicians into thinking that other species of the composite family were unlikely hay fever plants. Instead, Wodehouse insisted that the similar form of pollen grains within the family and the similar reactivity of hay fever patients

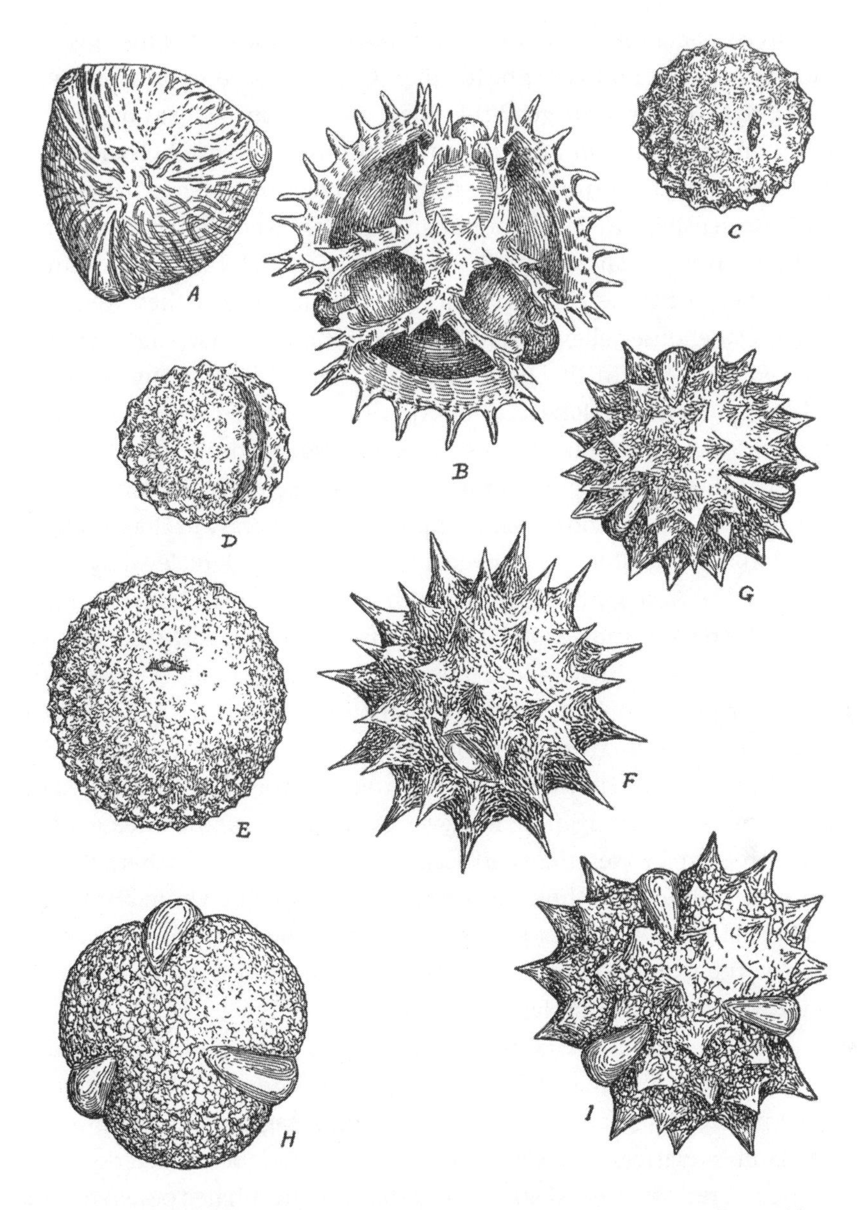

Figure 11. Pollen grains of quassia and composite families. The pollens of hay fever plants, including short ragweed (C), cocklebur (E), and sagebrush (H) provided research material for botanists such as Roger Wodehouse and a new means to study the evolutionary relationships of plant species. Roger Wodehouse, *Hay Fever Plants* (Waltham, Mass.: Chronica Botanica, 1945), 129.

to species of ragweed, cosmos, and sunflower demonstrated the close relationship of the ragweed tribe to other plants in the composite family.[24] Similarly, Wodehouse argued that the classification of Texas mountain cedar and red cedar in a genus separate from Port Orford cedar under the family Pinaceae went against immunological evidence. Hay fever patients sensitive to Texas mountain cedar showed similar sensitivity to Port Orford cedar, which was found only in California. Given that the pollens of these two different species reacted as if they were the same, Wodehouse believed that the trees' grouping into separate genuses marked "a defect in the botanical classification," especially since all other groups in the Pinaceae family, which included pines, firs, and spruces, were "notorious" for their lack of allergenic properties.[25]

Wodehouse was not the only botanist whose work on plant taxonomy contributed to and depended upon clinical allergy. To Harvey Monroe Hall, a botanist at the University of California, Berkeley, hay fever plants of the West, particularly the sagebrushes (Artemisia), offered abundant material for study, as well as a lucrative side income. Hired by Grant Selfridge, Hall conducted the first botanical and pollen survey of western hay fever plants in 1916. Selfridge traced his interest in hay fever to a Southern Pacific engineer from Los Angeles who had come to his clinic in search of relief from hay fever. Few reported cases of the illness existed in California. On investigation, however, Selfridge found the disease more widespread in the West than had been imagined. Among Southern Pacific workers, he found five hundred cases, two hundred of whom resided in California. Since hay fever presented an occupational hazard to railway men working outdoors on the tracks, the enterprising doctor convinced Southern Pacific to fund Hall's survey.

Traveling along the railroad's right-of-way through Utah, Nevada, Oregon, and California, Hall collected pollens from the most prevalent wind-pollinated species, including cockleburs, sagebrushes, mugworts, and lamb's-quarters. Selfridge, a hay fever sufferer, then tested these on himself. A real "twister," Selfridge told Hall after he inhaled pollen from a species of grass collected above Roseland, Nevada.[26] Hall himself found the work "very attractive." It is "like panning for gold," he wrote his family from the field, "and the pollen is more valuable than gold by weight." In his field journals, Hall commonly wrote down "gold" when he meant pollen. For mugwort pollen, he received more than $20 an ounce—the going rate for gold—from the pharmaceutical firm Parke-

Davis.[27] Other pollen collectors also commonly made the comparison to gold prospecting. But hay fever plants yielded Hall more than just pollen and additional income. His botanical survey resulted in a major publication on the ecological and evolutionary taxonomy of Artemisia, a group of wind-pollinated annual and perennial herbs and shrubs (including sagebrush and mugwort) that, next to ragweed and possibly the grasses, are the most important group of hay fever plants in the United States. It also earned him a prized research position at the Alpine Laboratory in Manitou Springs, Colorado, under the directorship of Frederic Clements, one of the leading plant ecologists of the day.[28]

In turning pollen into poison, allergists and patients transformed more than botanical knowledge. They forever altered the place of certain plants in American life. Pollen vaccines promised one form of hay fever prevention. Weed eradication promised another. At the U.S. Hay Fever Association convention in 1915, where Dr. Seymour Oppenheimer and Dr. Mark Gottlieb, physicians at Columbia University's Laboratory for Clinical Research, demonstrated the newly introduced technique of pollen vaccination, association secretary P. F. Jerome reported on anti-noxious weed laws up for legislative passage in ten states. The New Hampshire legislature rebuffed the association's efforts to pass a weed ordinance targeted at ragweed and other hay fever plants. But a model law passed successfully in Michigan—whose economy also depended on the hay fever tourist trade—and gave the association hope that other states would follow.[29] The passage of weed ordinances required more than legislative approval, however. It required the passage of a plant from the wilds of nature to the wastelands of civilization, a journey that had made ragweed public health enemy number one.

In the native landscape of North America, ragweed prospered in places of freshly disturbed soil—along riverbanks and on flood plains, deltas, and erosion gills and gullies. Botanists believed that South America was its place of origin. It then traveled along the foothills of the Andes, made its way through Central America, and settled in North America, where it established itself in patchy areas east of the Rockies long before the European invasion. To America's leading nineteenth-century botanist, Asa Gray, *Ambrosia artemisiifolia* was an "extremely variable weed" that flourished in "waste places" and flowered from July to September.[30] A relatively scarce plant until axe and plow opened up new habitat, ragweed spread across the North American landscape with civilization's advance. No stranger to humans, it followed railroad rights-of-way into the shipping yards, factory districts, dumps, and neglected neighborhoods of America's expanding industrial metropolises (Figure 12).

Ragweed found a comfortable home in the vacant lots and waste places of urban slums and tenement districts. These were places where few other plant species could survive. They were also areas of increasing importance to Progressive reform efforts, championed by public health officials, social workers, and civic organizations to improve the physical, moral, and social environment of the urban poor. Amid the rubble, refuse, and abandoned lots of lower-working-class neighborhoods, ragweed shared its home with other immigrants to the city, many of them from Southern and Eastern Europe, who had come to the United States in the late nineteenth century seeking their fortunes or at least a better quality of life.

Chicago's nineteenth ward was one such neighborhood. In 1889, Jane Addams and Ellen Gates Starr established Hull House in this southwest quarter. It was just north of the stockyards made famous a few years later by Upton Sinclair's novel *The Jungle,* which described how 250 miles of track moved 8-10 million animals per year into factories that produced not only food but also rivers of smoke and blood that darkened the sky and filled the air with an "elemental odor, raw and crude."[31] Here, Addams observed, "the streets are inexpressibly dirty, the number of schools inadequate, sanitary legislation unenforced, the street lighting bad, the paving miserable and altogether lacking in the alleys and smaller streets, and the stables foul beyond description,"[32] but thousands of Italian, German, Bohemian, Polish, and Russian immigrants struggled to gain a foothold in American society. This was the wasteland of industrial America, where the city dumped its garbage and its poor—an unhealthy place where sickness and death prevailed. Immigrant children, future generations of Americans, played in streets of mud mixed with dead and decaying matter. In their reform efforts, activists like Addams aimed to improve the living and working conditions of the urban poor. Hull House member Alice Hamilton went into the factories, investigating how workplace conditions exposed laborers to harmful levels of dust, lead, and other chemicals. Others, like Florence Kelley, took to the streets, knocking on the doors of tenements and boarding rooms to

Figure 12. Giant ragweed (*Ambrosia trifida*). Oren Durham, "The Pollen Harvest," *Economic Botany* 5 (1951): 217.

document and map the conditions of poverty, crowding, and unsanitary and decrepit housing in the neighborhood. Through the tireless work of these women reformers, a map of Chicago's urban ecology emerged that highlighted the relations among the physical environment, poverty, poor health, and social decay.

Transforming these landscapes of treeless streets and "unnatural stone" became one focus of women reformers eager to bring moral and social order to the urban environment.[33] Women's civic groups, believing that play on public streets contributed to vagrancy and juvenile delinquency, persuaded public officials to clear vacant lots of garbage and rubble and convert them into parks, playgrounds, and community gardens (Figure 13). In Chicago's nineteenth ward, the Relief and Aid Society employed homeless men (representing fourteen different nationalities) to clear vacant lots of "dead dogs, tin cans, wire springs, and all sorts of rubbish" and to fill in bad holes with cinder, ashes, and sand.[34] Once waste and filth were removed, order could be restored. Trees and shrubs were planted; swings, seesaws, and sandboxes were brought in; and playground supervisors were employed. Urban reformers believed that parks and playgrounds, breathing spaces for the urban poor, introduced a moral and healthful atmosphere into places where disease, crime, and poverty had grown.

A city's middle- and upper-class residents believed that vacant lots were places of waste, poverty, and filth. Rarely, if at all, were vacant lots—or the residents who lived nearby—considered by-products of the inefficiencies and injustices of industrial capitalism. To many, those who lived in the wastelands of the city were simply ill adapted to urban life. In the slums, plants and humans led a precarious existence. Insignificant and too often short-lived, both were considered (as noted above) urban weeds. "We all know," wrote physician Robert Hessler in a 1911 essay on weeds and diseases, "how large cities with a river front are infested by a class of people known as 'river rats,' a highly undesirable class, human weeds, so to speak." In Hessler's botanical and social survey of cities, the plants and human residents of "Shanty Town"—where waste prevailed—were nearly all foreign born. Some, Hessler suggested, lived there because of an inherently flawed nature. Others, if placed in a more sanitary environment, would become "desirable citizens," just as a cleanly and carefully tended orchard would yield healthy trees and bountiful fruit.[35] Cleaning up the ills of society by getting rid of vermin and people that

Figure 13. Vacant lots, such as this one in Chicago's nineteenth ward, were a target of urban reform, as well as a prime habitat for ragweed. B. Rosing, "Chicago's Unemployed Help Clean the City," *Charities and the Commons* 21 (1908): 50.

harbored disease was fundamental to the Progressive push to create a clean, healthy, and efficient urban order.

Ragweed's migration into the city, and particularly into city slums, made it, like certain other neighborhood transients, an "undesirable citizen." The plant did not, however, get its reputation as a "vegetable criminal" from simply hanging out in the seedy parts of town. Even though ragweed could "boast of being [100] percent American," it had characteristics that distinguished it from more appealing native plants.[36] In a national poll conducted in 1918 by the weekly magazine *The Independent,* goldenrod and columbine were selected as the two most favored candidates for the national flower. When a concerned reader wrote to *Science* that goldenrod accounted for 15 percent of hay fever symptoms, William Scheppegrell, a New Orleans laryngologist and president of the American Hay Fever Prevention Association, came to the plant's defense.

Goldenrod had gotten a bum rap, Scheppegrell said. In the northern, eastern, and southern states, the pollens of ragweeds (Ambrosiaceae) were the principal causes of fall hay fever. In the Pacific and Rocky Mountain states, the wormwoods (Artemisia) were the main culprits. Spring hay fever was largely caused by the pollens of grasses.[37] Moreover, there were specific reasons why goldenrod should not be included in the "rogue's gallery of the plant world."[38] All hay fever plants, Scheppegrell argued, displayed four characteristics that goldenrod did not share: "(1) they are wind-pollinated; (2) very numerous; (3) the flowers are inconspicuous, without bright color or scent; (4) the pollen is formed in great quantities." These were characteristics of "plants which occur as weeds in empty lots, neglected gardens, sidewalk and waste land generally."[39] Hay fever plants appeared wasteful, not simply because of the places where they thrived, but also because of their reproductive traits. They produced pollen far in excess of the need for survival into the next generation—just as the immigrant poor, American eugenicists argued at the time, outreproduced more respected citizens in society.

Ragweed's alleged lack of reproductive restraint made it the target of legislative action, not unlike Eastern and Southern Europeans, who suffered under the U.S. Immigration Restriction Act of 1924, designed to limit their entry into America. Despite ragweed's American heritage, it too became part of the "immigrant menace" that threatened the health of the nation. Some felt it deserved one fate: removal.

Although weed ordinances existed in many municipalities and states, some dating back to the 1860s, rarely were such bills enforced. In the early 1910s, the U.S. Hay Fever Association began lobbying its members to write to their local boards of health to encourage a more active policing of city weeds. Most anti-weed laws, however, applied strictly to weeds along highways and on private property deemed a nuisance to agriculture. Seeking a noxious weed law regulated not by committees of forestry, agriculture, or highways but by committees on public health, the association in 1915 drafted a sample bill modeled after Michigan's noxious weed law to be introduced in other state legislatures; the bill singled out ragweed, Canadian thistle, milkweed, wild carrots, oxeye daisies, goldenrod, and other noxious weeds that were believed to be aggravating to hay fever sufferers. Highway commissioners would be responsible for cutting such weeds along roads twice each season, while property owners within the city limits would be responsible for destroying weeds on their premises. Those who disregarded the ordinance would risk a lien on their property until municipal authorities did the removal and the city's expenses were paid. In New York, the association's lobbying efforts proved successful. In 1916, an amendment to New York State's noxious weed law was added that charged persons failing to cut certain hay fever weeds with a misdemeanor and a fine of not less than $5 and not more than $25 for each violation.[40]

The earliest, most systematic municipal weed eradication effort in the name of hay fever prevention took place in New Orleans in the summer of 1916. The principal organizer behind this campaign was William Scheppegrell, goldenrod's defender. Scheppegrell had established one of the first allergy clinics in the country at New Orleans Charity Hospital. There he began working closely with botanists, the U.S. Department of Agriculture, and the U.S. Public Health Service in extensive local and regional studies and surveys of hay-fever-causing plants found in North America. Scheppegrell argued that a physician would have more success treating hay fever "if, instead of limiting his attention to writing a prescription or injecting a vaccine, he investigate[d] the presence of hay-fever producing weeds in the neighborhood of the patient's residence or vocation." In May 1916, Scheppegrell boldly asserted to members of the Louisiana State Medical Association that the number of hay fever cases could be reduced by 50 percent if common and giant ragweed were eliminated from the city of New Orleans.[41]

Mapping the location of ragweed throughout New Orleans was the responsibility of a topographic committee that relied upon a variety of informants. In its treatment of hay fever, Scheppegrell's clinic furnished patients with a nine-block map of their neighborhood. On it they were to locate lots "infected with weeds." Considerable assistance was also furnished by the Women's Civic League, which appointed a committee to report on the location of vacant lots where ragweed grew in abundance. Through their active efforts in urban relief, from vacant lot gardening to school playgrounds, women knew the geography of the city in ways that proved invaluable to the weed eradication programs of New Orleans and other cities that later followed suit.[42]

Weeding itself was an activity closely associated with urban reform. To clear the streets and sidewalks of weeds in the outlying districts of New Orleans, twenty convicts were placed at the service of the American Hay Fever Prevention Association. They provided cheap labor, to be sure. But a moral conviction also prevailed in using prisoners to uproot undesirable plants. Such a form of work relief dated back to the late nineteenth century, when Detroit mayor Hazen S. Pingree introduced vacant lot gardening to improve the unhealthy conditions of lower-working-class neighborhoods. The industrial slowdown and large-scale unemployment following the economic depression of 1893–1894 had strained city relief efforts and fueled labor unrest. In response, Pingree proposed turning municipally owned and privately donated vacant land over to the poor and unemployed so that they could raise their own food. Pingree's Potato Patch scheme exceeded all expectations. In the first year, 945 families turned 430 acres of vacant land in the city into productive gardens that yielded 14,000 bushels of potatoes, in addition to beans, cucumbers, corn, tomatoes, and turnips. In that year an initial investment of $3,000 reaped an estimated $12,000 in produce. Within three years, twenty-five cities had adopted vacant lot cultivation as a form of physical and moral reform. To reformers, "fresh air and moderate exercise" greatly benefited the "physical and moral health" of those living where concrete, refuse, and squalor seemed to proliferate.[43] R. F. Powell, superintendent of the Philadelphia Vacant Lot Cultivation Association, described how a "helpless paralytic [man] pulling the weeds from among the little plants" was transformed into a healthy, productive citizen who, in a matter of five years, became manager of a nine-acre farm.[44]

Given this history, it is not surprising that when city officials looked to institute ragweed eradication programs, they often turned to the destitute of society. During the Great Depression, for example, New York City put fifteen hundred unemployed men to work through the Works Progress Administration to rid the urban landscape—132,600,000 square feet, to be exact—of ragweed (Figures 14 and 15). Chicago too recruited an "army of 1,350 men" from the city's shelters, with the help of Good Will Industries and the Chicago Women's Clubs, to wage "war for hay fever sufferers against one of the most prolific ragweed crops Chicago has known in years." In cleansing the city of a notorious weed, the destitute not only gained employment but also earned the appreciation and respect of hay fever sufferers thankful for their "humanitarian work" for society.[45]

Weed eradication added to civic pride and the city beautiful, but it did little to alleviate the ills of urban life, including hay fever. Vegetation

Figure 14. The "war against ragweed" fought by New York City Department of Sanitation workers. *Life,* 24 August 1942, 52.

Figure 15. Ridding the urban landscape of ragweed in New York City. *Life,* 24 August 1942, 52.

maps of cities provided physicians with detailed knowledge of the relative abundance of local hay fever plants. However, they gave no information on daily, monthly, and seasonal atmospheric pollen counts that might offer clues to the severity of the allergy season. Scheppegrell was the first to begin regular aerial sampling of atmospheric pollen. He exposed glass slides coated with glycerin for twenty-four hours outside the eighth floor of the Audubon office building in downtown New Orleans. But the doctor had still higher ambitions. In 1924, this aerial daredevil of the allergy world flew a biplane to an altitude of fifteen thousand feet. As he did so, he sampled the air for pollen at thousand-foot intervals. The presence of ragweed, marsh elder, and other pollens at altitudes upwards of six thousand feet led Scheppegrell to conclude that his local weed eradication efforts were all for naught. Without weed legislation at the state and federal levels, local efforts would have little impact, since it was clear that pollen readily traveled across municipal, county, and state lines.[46]

Scheppegrell's high-flying escapades caught media attention. But in the decade when Charles Lindbergh crossed the Atlantic, pollen was hardly the subject of daily weather reporting and conversations. Today, of course, it is. For that, we can thank one man: Oren Durham. Born in a sod house on the western edge of Kansas, Durham turned his boyhood interests in botany and the outdoors into a unique career. He had tried his hand at many jobs—nurse, carpenter, and tinsmith, among others—before he stumbled upon pollen collecting as a lucrative pastime. In 1913, Durham's uncle by marriage, Dr. R. Claude Lowdermilk, started using active immunization to treat hay fever patients in his Galena medical practice. It was just two years after Freeman and Noon had published their initial results of immunotherapy research.[47] Within a few years, Lowdermilk was no longer able to collect on his own the quantities of pollen he needed for the therapeutic treatment of his patients. Durham helped, gathering the pollen of ragweed, marsh elder, and other suspected hay fever plants during their blooming seasons. Durham originally pursued pollen collection as a side income. But by 1923 the price of pollen was $300 an ounce for certain plant species, more than fourteen times the price of gold. The profitability of this new occupation enabled Durham to leave his job at a photographic studio in Kansas City and take up pollen prospecting full time. That same year, William Duke, a Kansas City physician and early president of the American Association

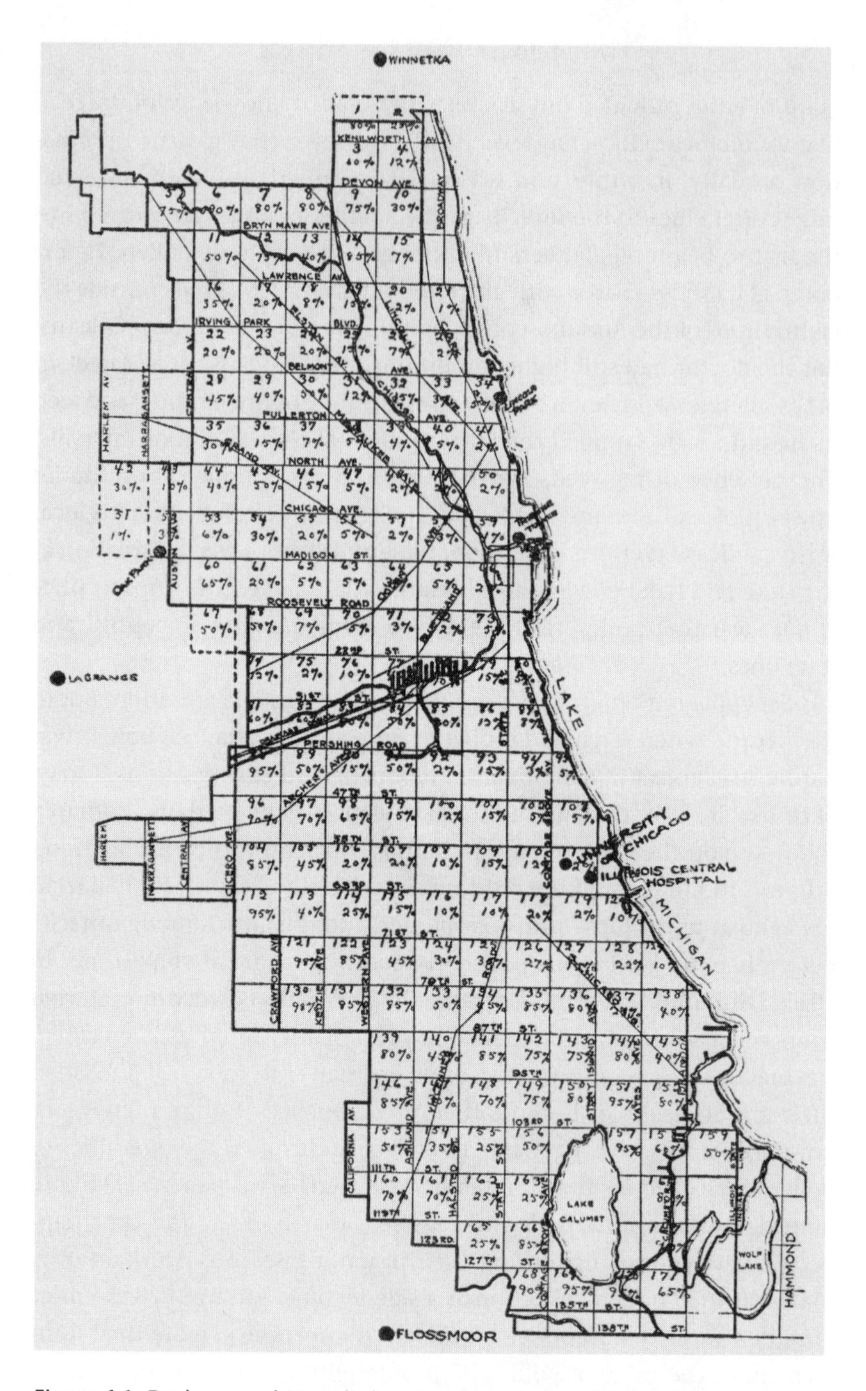

Figure 16. Durham and Koessler's extensive urban pollen survey of Chicago revealed the close correlation between urban "wasteland" and the presence of ragweed. Karl Koessler and O. C. Durham, "A System for an Intensive Pollen Survey," *Journal of the American Medical Association* 17 (1926): 1204.

for the Study of Allergy, hired Durham. For a fee of $3,000, he was to conduct an extensive vegetation survey of common wind-pollinated plants in Kansas City and the surrounding region. As part of this work, Durham analyzed daily air samples from the business district and suburbs of Kansas City for pollen content.[48] Duke tested the reactivity of hay fever patients in his clinic to the pollen of over fifty different plant species collected by Durham.

In 1925, as a "pollen gatherer" for the "allergenic products line" of Swan-Myers, Durham collaborated with Karl Koessler, who early on had expressed skepticism about hay fever vaccines, to develop what is perhaps the most extensive urban pollen survey ever undertaken. Utilizing the quadrat sampling method developed in plant ecology by another prairie native, Frederic Clements, Durham and Koessler divided Chicago into 171 square-mile blocks (Figure 16). (This wasn't the first time Clements's ecological methods had been applied to the study of hay fever. He and Harvey Monroe Hall had pioneered a more ecological approach to plant taxonomy based on Hall's survey of Western hay fever plants.) For each square, Durham and Koessler recorded the percentage of area occupied by vegetation, use of the area (for example, industrial or residential), and relative abundance of plants. The amount of wasteland and ragweed in the city correlated closely with the urban ecology of neglect. They determined that 38 percent of land in the city—or roughly forty thousand acres—scattered across industrial suburbs, railways, and canals, was overrun with weeds. This "crescent of neglected land" formed around the city an area "far weedier than the average farming district." It also roughly outlined urban neighborhoods, such as the nineteenth ward, where some of Chicago's most inhospitable living conditions predominated. Although middle- and upper-class residents in immaculately landscaped suburbs like Oak Park or Lake Forest lived distant from the weedy parts of town, they were hardly immune from the pollen scourge. Atmospheric pollen plates, exposed on a daily basis at various sampling sites in the city from June through September, revealed an abundance of pollen from grasses, ragweeds, and other hay fever plants. Durham and Koessler calculated that the ragweeds accounted for 65 percent of the total pollen load in Chicago and 80 percent during the fall hay fever season. The Windy City alone liberated hundreds of tons of ragweed pollen into the atmosphere each season—a bleak picture for hay fever sufferers too poor to escape or seek medical care.[49]

By 1928, Durham had collected data from pollen surveys made by twenty-eight allergy clinics in twenty-two cities across the United States. The lack of a standardized technique for air sampling, however, made comparative analysis difficult. To overcome the obstacles of private local studies, Durham enlisted the cooperation of the U.S. Weather Bureau to coordinate and standardize pollen sampling into a national atmospheric survey. Through its services, Durham distributed uniform sampling devices, materials, and instructions to local meteorologists in twenty-eight cities. At weekly intervals, he received a shipment of slides, which by 1936 totaled eighteen thousand. Durham personally identified and counted pollen from all species of the ragweed family on every slide. Pollen counts were correlated with meteorological factors, including temperature, sunshine, and rainfall, as well as the relative abundance of plant species and land-use patterns. In 1936, the number of sampling locations exceeded one hundred and extended from Winnipeg, Manitoba to Tampico, Mexico. The following year, the New York *World-Telegram* became one of the first newspapers to begin publishing in its weather section a daily pollen count, which is today a standard part of television weather forecasts.[50]

National pollen maps proved extremely useful to pharmaceutical companies such as Abbott Laboratories, where Durham was hired as chief botanist in 1930 to develop a market for pollen extracts. Although allergists expressed dismay in the 1930s about their reputation as "quacks" who made a "good living scratching the skin and pushing the needle," the number of allergy sufferers was on the rise.[51] In less than two decades, estimates of the number of hay fever sufferers had doubled from 1.5 to 3 percent of the U.S. population. Hay fever was no longer a disease of the wealthy. By the 1930s, it had become relatively common in urban America, where the medical practice of allergy was also on the rise. In twenty years, the Western Society for the Study of Asthma, Hay Fever, and Allergic Diseases and the Society for the Study of Asthma and Allied Disorders had grown from founding memberships of less than 20 in 1923 to over 272 when the two organizations merged in 1943 into the American Academy of Allergy.[52]

Allergy was a growing market. Pollen was both poison and profit. Drug companies capitalized on increased demand for pollen extracts by furnishing physicians throughout the United States with pollen maps and data organized by place and season. Abbott Laboratories reminded doctors that "the variety and irregular distribution of hay fever flora"

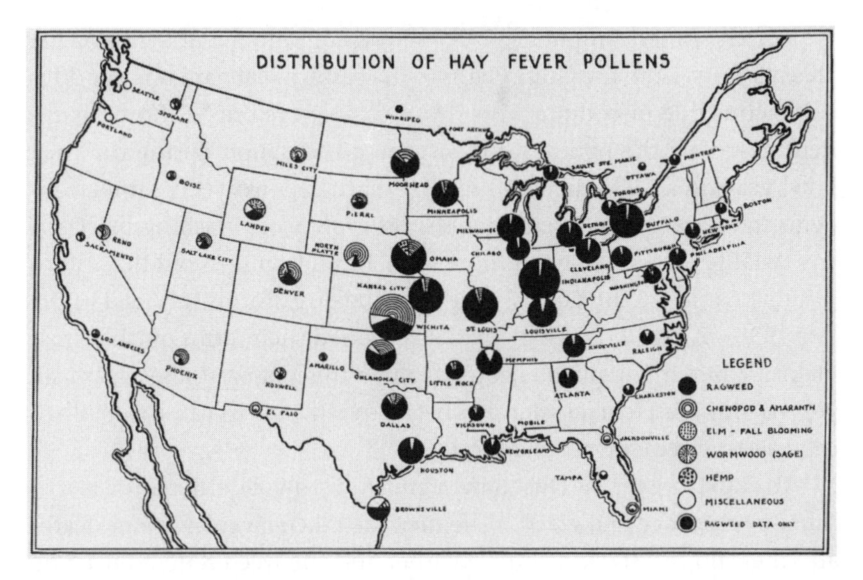

Figure 17. Pollen maps of the United States, such as this one from the 1930s, placed ragweed firmly within the ecological crisis of the Dust Bowl and many, such as Roger Wodehouse, attributed its spread to the immorality of thoughtlessness and greed. Reprinted from O. C. Durham, "The Pollen Content of the Air in North America," *Journal of Allergy* 6 (1935): 129; with permission from the American Academy of Allergy, Asthma, and Immunology.

meant no "two physicians working in different cities" required "exactly the same pollens for local testing." Then to fill the need, Abbott offered a "comprehensive regional skin test" service that gave advice on the "practical selection of local pollen tests for any district."[53] Similarly, Lederle included in its brochure on pollen antigens a booklet of known hay fever plants in the New England, Eastern, Middle Western, Rocky Mountain, and Southwestern districts, along with illustrations and times of bloom.[54]

Durham's pollen maps did more than increase Abbott's sales. They offered a panoramic picture, an ecological view, of America's allergic landscape. Social surveys had guided the mapping of ragweed's and other plants' local, urban environments. Plant ecology shaped understanding of these plants on a regional scale. Pollen maps of the United States in Durham's popular 1936 book, *Your Hay Fever,* with the largest circles of ragweed pollen dominating the corn and wheat belt of the Midwest, placed ragweed firmly within the agricultural crisis then gripping the nation (Figure 17).

For the American public, 275,000 tons of ragweed pollen blowing "like a storm" from the Great Plains to the eastern seaboard was not difficult to imagine or comprehend. "An invisible cyclone" is how George Kent described the pollen storm sweeping the nation during the Dust Bowl years in a 1936 *American Magazine* article. Two years earlier, residents in eastern cities, including Boston, New York, Washington, D.C., and Atlanta, had watched as more than 12 million tons of Midwestern dirt had darkened the skies. By the time the dust storms subsided in the late 1930s, 2.5 million people had abandoned their farms on the Great Plains. While many plains people blamed this national disaster on six years of drought, federal scientists began to attribute its cause to ecologically destructive land-use practices.[55]

To Paul Horgan, the Dust Bowl signified a struggle with nature lost to human arrogance and greed. "The men are taken away by life or death; and their houses stay until the weather and the weed render dust," he wrote in his 1936 book, *The Return of the Weed*. "A government report has called them, these abandoned places of human passage, visible evidences of failure. They are monuments. They lie ruined, as memories of audacities long-gone, and poor judgements; the almost inscrutable remains of aspiration wedded to tragedy."[56]

Horgan was not the only one to see a moral lesson. Roger Wodehouse, by that time scientific director of the Hayfever Laboratory at Arlington Chemical, attributed the increase in ragweed and hay fever to "man's recent land abuses." By studying pollen accumulated in layers of peat built up over eons of time, Wodehouse found that the "enormous abundance of ragweed occurring in America" was a "comparatively recent development and undoubtedly the result of human activity."[57] Before European settlement, ragweed had been virtually absent in the pollen fossil record. Other physicians and botanists expressed similar dismay at how changing patterns of land use had altered the vegetation of particular areas and contributed to allergy's rise. The common use in Colorado gardens of summer cypress, an important cause of hay fever in the Rocky Mountain region, turned into an ecological and medical nightmare after the plant spread into the wild.[58] "Hay fever," Wodehouse told *Natural History* readers in 1939, "is nature's reply to man's destructive and wasteful exploitation of natural resources just as much as is soil erosion, wind erosion and floods. It is less spectacular than the great gullies carved out of hillsides by running water or the disastrous dust storms that bury farm buildings and move whole farms into the next state, or the floods that sweep away bridges. These are nature's answer in her boisterous mood. In her more subtle mood the answer is hayfever. And so softly it comes that few of us ever suspect that it is the answer to our thoughtlessness or greed."[59]

Among plant ecologists confronted with the Dust Bowl, a catastrophe of immense ecological and economic proportions, weeds were more valued than despised. Ragweed became a "pioneer," not an "invader," a symptom of human exploitation of the land. "Weeds, like wild-eyed anarchists," wrote ecologist Paul Sears, looking out on the blown-out, baked land in Oklahoma, "are the symptoms, not the real cause of a disturbed order."[60] In the minds of plant ecologists, western hay fever plants such as prairie sagewort (*Artemisia frigida*) were signs of overgrazing on the Great Plains. Ecologically unsound and unsustainable land-use practices had upset the balance of nature. To ecologists, the tall-grass prairie of the Great Plains represented a climax community, "a stable and balanced society of plants" that had evolved over time and was "best suited to the soil and climate" of the region. Weeds indicated that not all was well on the prairie. But weeds were also important pioneers in regenerating the health of the land. Their presence signified an early stage in the development of a climax community, a process ecologists called succession. Native weed species gained a firm foothold in barren soil and prepared the ground for other plants. "Ragweeds . . . perform a useful service," Wodehouse noted, "in holding the soil against wind and water erosion until it is taken over by other more permanent plants." In the art of "land doctoring," weeds were both a diagnostic tool of plant ecologists and nature's first therapy in land restoration.[61]

Ragweed's identity as a native pioneer species and its presence in the nation's upper atmosphere existed at odds with its place in the city, where its immigrant status and prolific reproductive habits made it the target of removal. Oren Durham looked skeptically on municipal weed eradication programs, as did Wodehouse. Given the prodigious quantities of ragweed pollen in the upper atmosphere, Durham argued that local efforts, no matter how widespread and successful, would not have much effect. Wodehouse agreed: "Ragweed . . . will be here . . . after the human race has ceased to trouble this planet, faithfully doing its job of soil conservation." Citizens needed to follow nature's lead and practice sound soil conservation if ragweed's presence in the city was to diminish. Appropriate plantings in cleaned-up vacant lots would be far more effective in the long run, Wodehouse argued, than the usual "root-and-burn" method, which attacked the "symptom, not the malady itself." Pulling ragweed up by the roots only created more favorable conditions for the plant the following year. The newly disturbed soil offered an ideal environment for the germination of ragweed seeds, which could lie dormant in the ground and remain viable for forty years. In Wodehouse's urban ecology, ragweed may have been a "vagrant riffraff of the plant world." Nevertheless, it held value as nature's instructor, saving billions of dollars in soil protected from erosion. Ragweed was nature's ecological pioneer and subtle revenge *against* the ravages of human civilization.[62] Only when humans became responsible stewards of the land, Wodehouse suggested, could they prevent the spread of ragweed and allergy in America.

While botanists and plant ecologists appealed to a conservation ethic in the prevention and control of hay fever during the Great Depression, an engineering ethos dominated ragweed control measures after World War II. Wodehouse advised city health officials and sanitary engineers in the 1940s that treating the environment instead of the patient was the only way that hay fever would ever be cured. And they listened. But where Wodehouse saw in allergy's rise a cautionary tale of the need for humility and respect for nature's economy, public health departments, armed with the wondrous technological results of wartime research, entered the postwar era with renewed confidence in their ability to control, if not eliminate, hay fever. Scientists had mastered the secrets of the atom and harnessed its energies. How could humans be conquered by a simple weed?

At its 1947 meeting the American Public Health Association (APHA), considered the many smelters, cement factories, oil refineries, and weapons laboratories that discharged smoke, dust, radioactive isotopes, obnoxious odors, and other hazards into the atmosphere as among the industrial operations potentially harmful to the public's health. Yet it was a simple weed—ragweed—that stood out as a serious health problem.[63] Next to newly unleashed radioactive isotopes that came into the world with the advent of the atomic bomb, public health officials agreed, pollen was the most serious health threat of known atmospheric contaminants. "Pollen factories," each "operating without Federal permission," annually produced an estimated one million tons of "toxic dust,"

Figure 18. By the 1950s, ragweed pollen had become public health enemy number one. *Hay Fever Bulletin* 6 (1955).

275,000 tons of which made its way into the air. Nearly 93 percent of the population of the United States was exposed to this "atmospheric contamination," produced in the ragweed belt east of the Rocky Mountains.[64] "The control of atmospheric pollution," the APHA's Committee on Air Pollution stated, "is an administrative and technical problem that can be solved by engineering means."[65] Ragweed, a source of air pollution threatening the public's health, had suddenly become entwined with industrial regulation and control (Figure 18).

An industrial plant like ragweed required industrial-strength control measures. Luckily for sanitation engineers determined to eliminate one of the city's most notorious polluters, the nation's military-industrial complex had produced an ideal weapon of destruction: the herbicide 2, 4-D. The development of 2, 4-D grew out of interwar research in plant physiology on growth hormones that became part of a top-secret wartime project of the National Academy of Sciences on biological and chemical warfare. After the war, a major dispute and a prolonged legal battle erupted between the American Chemical Paint Company and Dow Chemical over who had discovered that 2, 4-D was an effective herbicide. In spite of the dispute, the chemical quickly made its way into the marketplace. Sold under various brand names (including Weedone, Weedex, Weed-Be-Gone, and Weed-No-More), 2, 4-D was initially met with great enthusiasm by suburbanites and golf course owners who sought to keep their lawns and fairways free of broadleaf weeds. Within a few years, agricultural scientists convinced farmers of its benefits. In the field of public health, 2, 4-D spraying campaigns against ragweed threatened to surpass municipal applications of DDT for mosquito control. By 1949, it was estimated that 20 million pounds of the weed killer had been produced.[66]

Operation Ragweed was "fought in the lots and backyards of the five boroughs" of New York City and was among the earliest, largest, and most sustained "battles of municipal forces attempting to vanquish a sore spot of nature." In the summer of 1946, the city's Health Department launched an all-out chemical war on ragweed. The $85,000 it cost seemed a small price to pay to help the estimated four hundred thousand city residents afflicted with hay fever, who had little more than handkerchiefs as a defense against "the sneak-enemy attack by pollen."[67] Federal, state, and municipal agencies converted mosquito control units, road-oiling equipment, street flushers, and tree sprayers into vehicles armed for combat. As many as thirty-three trucks equipped with multiple-nozzle guns sprayed 850,000 gallons of 2, 4-D on three thousand acres of public properties, roads, sidewalks, and privately owned vacant land where ragweed was found. Local police departments undertook ragweed reconnaissance missions. Their investigations helped supply spraying crews with maps that contained information on the location and abundance of the fearless floral foe. The involvement of the police in vegetation mapping only reinforced the "moral space" vacant lots occupied within the urban ecology of reform.

New York City's health department also launched a widespread public education campaign that enlisted the aid of civic groups, chambers of commerce, hay fever and ragweed societies, and educational institutions. Boy Scouts took to the streets in a door-to-door campaign to distribute illustrated pamphlets and posters that reminded law-abiding citizens of their legal obligation to destroy ragweed.[68] The spraying of 2, 4-D was promoted as the most effective eradication method. Although

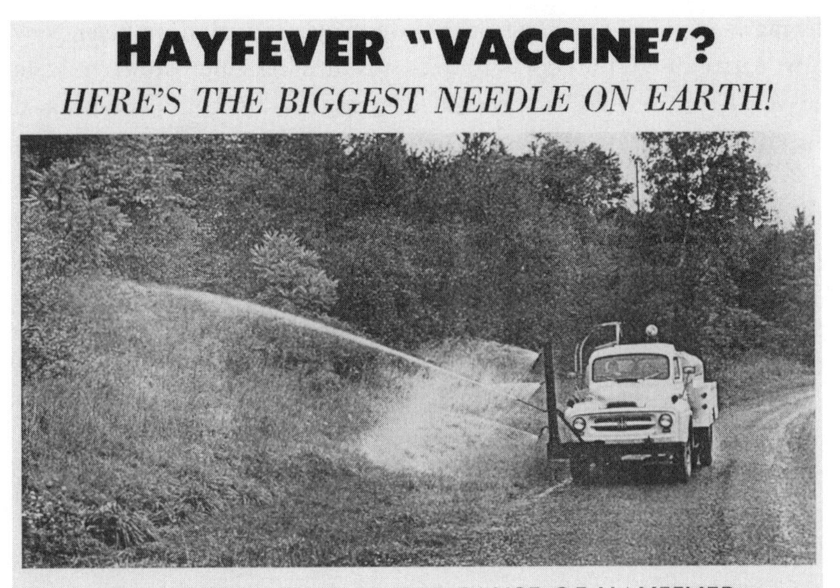

Figure 19. Herbicides like 2, 4-D, used in municipal ragweed control programs, signified postwar American optimism about the complete technological control of nature. *Hay Fever Bulletin* 6 (1955): 2.

an editorial in the *Journal of the American Medical Association* ridiculed the belief that any community could permanently rid itself of ragweed where the plant had already become well established, Health Commissioner Israel Weinstein remained determined. We will fight "for as many years as necessary," Weinstein proclaimed, on as wide a front as possible, until ragweed is eliminated from the urban landscape.[69]

The first year of Operation Ragweed exceeded all expectations. Atmospheric pollen content dropped to almost half that of the previous year. Weinstein brushed aside suggestions that the season's unusual rainfall had played a role and credited instead the destruction of about one-third of the city's ragweed as an "important factor in relieving the distress of hay fever." Chicago, Detroit, Jersey City, Washington, D.C., and many other smaller towns, communities, and counties followed New York City's lead. By the early 1950s, responding to municipal demand, manufacturers such as McMahon Brothers designed and sold spray units that could be driven and operated by one person to cover "infected" areas of ragweed with this new "hay fever vaccine" (Figure 19). Plagued by ragweed pollen blowing from the west side of the Hudson River, New York City enlisted the cooperation of surrounding suburbs and the state of New Jersey in what had become to engineers—but not to botanists—a surprising interstate problem in pollution control. By 1955, more than 150 communities in New Jersey had adopted ragweed-spraying campaigns. Such programs helped push the sale of 2, 4-D to more than $50 million that year.[70]

Nine years and 8 million gallons of herbicide later, Philip Gorlin, supervisor of New York City's Department of Air Pollution Control, claimed that Operation Ragweed was a model of what communities could do in hay fever prevention and control. America's enthusiastic embrace in the 1950s of DuPont's slogan, "Better living through chemistry," continued to generate public support for ragweed-spraying campaigns. In 1957, the year before New York City abandoned Operation Ragweed because it was dismayed by its costs and questionable success, the Oregon state legislature enacted a Ragweed Control Law. The law declared common and giant ragweed to be a public nuisance that threatened the "health and welfare of the people of the state." Exercising the "police power of the State" to protect its people "from the injurious consequence of ragweed pollen" and "to prevent further spread of this obnoxious weed," the act mandated the Department of Agriculture to detect, control, and destroy ragweed through the use of herbicides on public and private lands. Herbicides had become the would-be technological quick fix in America's war on allergy.[71]

The success of spraying campaigns such as Operation Ragweed in eliminating hay fever was questionable at best, however. And the toxins herbicides themselves introduced were hardly offset by whatever temporary benefits accrued from a minimal atmospheric reduction in pollen loads. In selling Operation Ragweed, Gorlin claimed that ragweed occupied only half the acreage it once had in New York City and that pollen counts had also dropped. But Gorlin played fast and loose with the evidence in his advocacy of engineering principles applied to public health. Matthew Walzer and Bernard Siegel, allergy specialists at Brooklyn's Jewish Hospital, had been measuring the effectiveness of New York City's ragweed eradication campaign since its inception. From pollen surveys taken at thirty sampling stations in New York City and its environs over a nine-year period, the doctors found no "consistent decline in New York City's pollen index." What Walzer and Siegel found instead was far more intriguing. Since 1946, weather patterns had blessed the city with an abundance of ocean breezes free of pollen. Prevailing winds from the west and southwest normally accounted for more than half of New York City's pollen load. But during the years in which Operation Ragweed had been under way, these winds had blown with less frequency and force than in previous decades. Climate patterns, rather than chemicals, deserved credit for the fewer sneezes in the city. So too did a building boom that reduced the amount of vacant land in Brooklyn by 60 percent from 1946 to 1954. In the postwar period, climate and land use had proved far more influential factors than chemicals in hay fever prevention and control. Both played a central role in the ecological understanding of allergic disease that had first been advocated by Wodehouse and others in combating allergy's initial rise.[72]

In *Silent Spring,* Rachel Carson pointed to weed eradication campaigns as an example of the "way efforts to control nature sometimes boomerang." Blanket spraying along roadsides created open barren areas, habitats conducive to ragweed's spread. Public health experts and sanitary engineers had failed to see allergic disease as a "problem of ecology, of interdependence of relationships." Their solution—chemical warfare—was symptomatic of an "era of specialists, each of whom sees his own problem and is unaware of or intolerant of the larger frame into which it fits." Cancer, genetic deformations, and dying songbirds were just a few of the specters Carson evoked to demonstrate the price humans paid for their lack of "humility before the vast forces with which they tamper." Just as Wodehouse saw in ragweed's proliferation nature's reply to human arrogance and greed, so too did Carson emphasize the capacity of life to strike back in "unexpected ways." Her book, published in 1962, evoked the ire of the chemical industry and forced American consumers to see their relationships to chemicals and the natural world in a new way.[73]

When ragweed moved from the wilds to the metropolis, it had not stepped outside its place in nature. Humans had, by ignoring how their actions shaped the urban ecology of plants and people. The herbicide 2, 4-D treated the symptoms, not the cause, of hay fever. It offered Americans an easy alternative to confronting the ecological relationships—cultural, social, and physical—that contributed to the growth and spread of ragweed and, as a consequence, allergic disease.

At the time of 2, 4-D's release, chemical manufacturers added another weapon to the arsenal in America's combat with hay fever: antihistamines, which first became available as prescription drugs in the mid-1940s. Over fifty years later, in 2001, $4.3 billion in annual sales had placed antihistamines among the top ten best-selling drugs in the United States.[74] Like 2, 4-D, these chemicals work by controlling the environment, but from within, not without. By altering chemical relationships within the body, pharmaceuticals lift the ecological constraints of the external world. In an ad for the "new and improved" antihistamine Claritin, allergy sufferers are now told they can use the product to "get back to nature," which is represented by a beautiful meadow of wildflowers in hyper-real colors. Through antihistamines, pollen is made pure once again. After a long and arduous journey, ragweed, through the technological wonders of biomedicine, has returned to its benign place in nature.

Herbicides may be easy to use, and antihistamines may be easy pills to swallow. But allergy sufferers are no more immune from environmental change than when nature offered the only refuge. As atmospheric carbon dioxide levels double over the next century, as they are predicted to do without a dramatic curtailing of global warming, scientists forecast that ragweed pollen loads could increase by more than 60 percent. Like many plants, ragweed is carbon-limited—that is, its growth and reproduction are dependent on concentration levels of carbon dioxide in the atmosphere. As CO_2 levels increase, as has been the case since the Industrial Revolution, they stimulate a ragweed plant's ability to grow and reproduce. So-called urban heat-island effects—the results of a city's tendency to be significantly warmer than the surrounding area—are linked to both increased carbon dioxide levels and air temperatures in and around cities. Already researchers with the U.S. Department of Agriculture have found that urban heat-island effects greatly influence seasonal pollen production of hay fever plants such as ragweed. In Baltimore, for example, ragweed plants in the city produce ten times more pollen than their rural counterparts. As William Schlesinger, dean of the Nicholas School of Environmental and Earth Sciences at Duke University, observed, "I see this as a potentially rather large health problem." It is a problem that all the herbicides and over-the-counter drugs in the world will not solve. It is a problem of ecology and disease.[75]

CHAPTER

3

The Last Resorts

Sought gold here, but found a greater treasure—good health.

—George Ragan, 1873

June Hewitt had come to dread the weekly eight-mile drive south from her Arizona ranch in Sulphur Springs Valley, east of the Mule Mountains, to Douglas, a town of twelve thousand residents just north of the Mexican border. She would have gone less often, but her truck could haul only a week's worth of feed for her horses and other ranch animals. Every time she drove past the sprawling, eighty-year-old Phelps Dodge copper smelter complex, around which Douglas was built, she felt a rawness in her throat. She knew an asthma attack would soon follow.[1]

Hewitt had suffered few health problems before moving to Sulphur Springs Valley in 1983. But as the five-hundred-foot-high smelter stacks spewed smoke that carried sulfur dioxide and other toxic airborne particulates as far north as the Rocky Mountains, she noticed a precipitous decline in her ability to perform ranch chores or household tasks (Figure 20). Every ten days or so, pollution rolled in "like a London fog over the bushes." When Hewitt could no longer "see the mountains anywhere around," she was in the middle of a toxic cloud that would send her to the hospital emergency room. Her physician, Dr. John Abbott, prescribed powerful steroids, the latest generation of asthma drugs, after she was hospitalized during a severe asthma attack in the fall of 1985. But even the wonders of biomedicine could not insulate Hewitt from the harmful effects of a degraded environment; as acid-forming sulfur

Figure 20. Sulfur dioxide plumes from the five-hundred-foot-high smelter stacks near Douglas, Arizona, jeopardized the health of asthmatics, not only in southern Arizona, but even in Denver, almost one thousand miles away. Douglas Smelter, date unknown. Phelps Dodge Collection. Courtesy of the Bisbee Mining and Historical Museum.

dioxide settled in her lungs, it triggered an inflammation of bronchial airways that left her gasping for breath. Abbott recommended, as a last resort, that Hewitt leave the area.[2]

A century earlier, physicians often gave this same advice to asthmatics who were left with little hope. Hay fever sufferers comforted themselves knowing that a carefully planned holiday or a change of seasons would likely bring relief from their affliction. But to the serious asthmatic, changing place to escape illness was often a permanent move, not a seasonal sojourn of summer fun. In the late nineteenth and well into the twentieth century, patients with intractable asthma were urged to pack up and move to the dry climate of Arizona or the cool mountain air of the Rockies. Western towns such as Tucson and Denver were built upon the rich natural resources of desert and mountain regions that attracted settlers seeking wealth—or health. The drive for progress made both cities into metropolitan centers of the West. But along the way, business and civic leaders and individual citizens made decisions that valued economic gains over the healthful properties of these regions and the health of residents.

Even if Hewitt had wanted to follow her doctor's advice, there was no obvious place to go. Denver had been a popular destination for asthma sufferers up through World War II. But by the 1980s, the combined output of 2,830 tons of sulfur dioxide per day from the Douglas smelter and two other uncontrolled smelters just across the border in Mexico—together the largest source of sulfur dioxide pollution in the United States and Canada—added to the deteriorating air quality of Denver, almost one thousand miles away, and to the increasing amounts of acid rain falling in the Rocky Mountains (Figure 21). The lives of both people and wildlife in these western regions were being endangered.

A western showdown was brewing in the plume clouds on the horizon. June Hewitt, Joyce MacKenzie Stillwater, and other asthmatics like them, some of whom had moved to the area in search of relief from their illness, believed the smelter was violating their "right to an uncontaminated air supply." Asthma sufferers, residents, and local physicians organized to form the Group against Smelter Pollution (GASP), an independent, grassroots, community-action organization whose goal was to reduce allowable emissions of sulfur dioxide to levels that would not seriously jeopardize the health of an estimated ten thousand asthmatics living in southern Arizona's Cochise, Pinal, and Gila Counties.[3]

GASP faced formidable foes. Phelps Dodge was a powerful Fortune 500 company. Moreover, an experienced gunfighter of Hollywood westerns, Ronald Reagan, occupied the Oval Office, and he was no friend of the environment. Under Reagan's administration, the Environmental Protection Agency (EPA) allowed smelters to avoid compliance with the National Ambient Air Quality Standards set forth by the 1970 Clean Air Act through a special permit known as the Non-Ferrous Smelter Order. In June 1985, when Bruce Babbitt, the rangy Arizona governor and ardent defender of western public lands, set foot in Washington and testified before the Health and Environment Subcommittee of the Congressional Committee on Energy and Commerce, what began as a seemingly local issue escalated into a symbolic fight over the dismantling of federal environmental regulations by the Reagan administration and the preservation of western lands. "Massive uncontrolled discharge[s]" by the Phelps Dodge smelter and the Nacozari and Cananea smelters in Sonora, Mexico, Babbitt told members of Congress, posed a "threat not only to

Figure 21. This cartoon from the *Phoenix Gazette* captured how a seemingly local environmental issue over smelter pollution had escalated into a national legal battle that was fought by GASP, the Environmental Defense Fund, and seven states to force the EPA to strengthen air quality standards under the Reagan administration. Courtesy of BORO, *Phoenix Gazette.*

public health in the U.S. but also to the pristine innermountain [*sic*] lakes in the West which lie directly on the northern wind currents. . . . If we follow the Administration's current course there will be no controls until 1988, at the earliest, and the residents of the Innermountain [*sic*] West will helplessly watch a million tons a year of sulfur scatter over our cities, lakes, and mountains." It was, Babbitt argued, "the gravest threat to public health and public resources" in the history of the West. Without federal action, Babbitt feared, "our health and natural resources [will] disappear" into the area that was home to three smelters and was becoming known as the "Gray Triangle."[4]

Two months later, Joyce MacKenzie Stillwater, GASP, and the Environmental Defense Fund (EDF) filed a petition with the EPA on behalf of those suffering from asthma to prohibit the continued operation of the Phelps Dodge smelter. Hewitt, Stillwater, and two hundred other allergy and asthma sufferers offered living proof of the dangers to Douglas residents of exposure to sulfur dioxide. EPA administrators had the authority to file suit on behalf of the United States to restrain a person causing or contributing to pollution that presented "an imminent and substantial endangerment to the health of persons."[5] When the EPA failed to respond, the EDF joined three other environmental groups, the attorneys general of seven states, and GASP to file suit in December 1985 against the EPA to force it to strengthen air quality standards for pollutants responsible for acid rain and the health problems of Arizona asthmatics. After a legal battle, Phelps Dodge closed the Douglas smelter in July 1986. (Smelting technology at the plant had not been renovated since 1913, not quite a decade after the smelter was first built to reduce ore from the nearby Bisbee Copper Queen mine, one of the richest copper finds on earth.)[6] GASP and environmental groups such as Earth First! celebrated, claiming that the people of Arizona, Mexico, and those living close to the smelter enjoyed "easy breathing and spectacular vistas they hardly knew existed."[7] But the victory was a foregone conclusion. Phelps Dodge, having recognized that it would be financially unable to bring its Morenci and Douglas smelter plants into federal compliance, had begun shifting its production from Arizona to New Mexico in the mid-1980s.

The battle over the West's natural resources, which pitted the needs of public health against those of development and progress, did not begin with asthmatics, physicians, environmentalists, or state leaders challenging federal environmental regulations and the interests of big business in the 1980s. Rather, it began a century before. In the late nineteenth century, civic boosters, business entrepreneurs, financiers, and physicians sought to bring civilization to the western frontier. They did so by promoting the natural resources of the Arizona and Colorado Territories that would enhance wealth, health, and pleasure. As a result, the railroads brought both prospectors and health seekers in droves to these regions—specifically to Tucson and Denver. Few could have envisioned that this tale of two cities built upon mining and health would end in a sulfurous haze that clouded national park vistas on the Colorado Plateau and threatened to turn what were once landscapes of hope and prosperity into landscapes of despair.[8]

It was the railroad, not acid rain, that first linked Tucson and Denver. In the expansion of the West, this engine of progress transformed frontier towns into bustling metropolises. Capital, merchandise, and people flowed from the East, and precious ores moved from the mining districts to industrial centers as the steam horse opened up rapid corridors of commerce. In 1870, the Denver Pacific Railroad first linked Denver to the transcontinental railway through the northern junction at Cheyenne, Wyoming. In the next ten years, six additional railroad lines and thirty thousand new residents helped establish the Mile-High City as the region's commercial center. By 1880, Denver, Colorado's capital city, had four national and two state banks with aggregate deposits of more than $6 million, a branch of the U.S. mint, and an annual eastern exchange of $65 million. It was an economy built not only on the region's rich mineral resources—an estimated $224 million in gold and $541 million in silver came from the surrounding mountains and streams between 1870 and 1900—but also on the climate, air, and sunshine of the region. The latter were sold and delivered by civic boosters, physicians, and railroads to health seekers looking for relief, if not a cure, from respiratory illnesses such as asthma and consumption.[9]

By 1890, an estimated 30,000 invalids—the majority of whom were asthmatics and consumptives—had come to Denver seeking their fortune in health. They made up almost a third of the city's population. Over the next forty years, the number and percentage of Denver's health seekers grew. By 1920, an estimated 40 percent of the city's 250,000 residents—"lungers," as they were known—had come in search of health.[10] Denver, the Queen City of the Plains, became the place of escape for the consumptive and the intractable asthmatic. And while it was the first western locale to capitalize in a big way on the health benefits of its surrounding nature, it was not the only one. When the Southern Pacific extended its reach to Tucson in 1880, the little Arizona outpost followed Denver's lead, turning a land of sunshine, dry air, and rich mineral resources into a metropolitan center of prosperity, health, and commercial progress.

Travel to the West was expensive in the decade following the completion of the transcontinental Union Pacific Railroad in 1869. A round-trip ticket from New York to Sacramento in the luxurious comfort of a

Pullman car cost $300 in 1872. Meals, hotels, carriages, and horses were extra. As a result, those who first came to Denver in search of health and pleasure were largely of the professional class. They were also the kind of clientele the city's powerful elite hoped to attract in building a financial center of the West.

The boom-and-bust cycle of mining—with its influence of ruffian prospectors, saloons, and bordellos—gave some business leaders pause. But they looked favorably on the treasure of health to be found in the Rocky Mountains: it attracted a refined class, was less prone to market fluctuations, and was built on a seemingly inexhaustible resource. Unlike the prices for silver or gold, Colorado's climate was predictable, at least in the 1870s. The Passenger Department of the Denver and Rio Grande Railway assured prospective health seekers that they could be guaranteed 302 days of sunshine in a year.[11] Denverites knew that guarantees could not be offered for mining investments, especially after stock in Colorado mining plummeted 71 percent following a national financial panic in 1873. Business organizations like the Denver Manufacturing and Mercantile Bureau promoted mining as the key to the city's financial future. But other town leaders expressed caution, pointing out that minerals were just one of many natural resources with which the territory was blessed. When the Leadville silver boom in 1877 created a "momentary rush of mining excitement," the editor of the *Rocky Mountain News* admonished Denver residents for forgetting the "very great importance of the invalid patronage." The newspaper reminded its readers that "thousands of families afflicted with consumption or asthma" came to Denver, disbursing "large sums of money, and [they could] be likened to permanently improving gold mines within our limits, in which every citizen shares the profits." If the city was to continue to attract wealthy health seekers from the East, however, the *Rocky Mountain News* urged, modern sewers, cement sidewalks, and a hydrant system were needed to keep the streets free from dust, which had become an increasing drawback to the city's "numberless invalids." "All does not depend on the mines," the newspaper observed.[12]

In attracting wealthy health seekers and promoting the salubrious environment of the West, railroads and town boosters both solicited and welcomed the praise and testimonials of nineteenth-century celebrities who espoused the region's health-restoring qualities. A weary passenger, glancing through the *Union Pacific Tourist* (one of a growing number of popular, informative, and finely engraved travel guides written for the railway traveler) could find, for example, humor and comfort in a quip by America's master showman and humbug artist P. T. Barnum, who joked that Coloradoans were the "most disappointed people [he] ever saw. Two-thirds of them come here to die and they can't do it."[13] Some of America's most prominent writers were happy to advertise the region's therapeutic benefits for asthma and consumption. America's leading female journalist in the late 1800s, Sarah Jane Lippincott, wrote to her *New York Times* readers while making a railway journey on the Union Pacific Railroad in 1871 that the air of Denver throbbed with "pulses of a new life" and that she did "not believe there is out of Heaven such a place as the mountain land of Colorado" for confirmed asthmatics like herself. Such testimonials proved the best of promotional copy and attracted countless others, who likewise hoped to find a new life in new lands (Figure 22).[14]

Perhaps no American literary talent did more to promote the virtues of Colorado and other health-restoring regions than Helen Hunt Jackson. Her seasonal wanderings in search of relief from hay fever and her accompanying travel writings had helped transform the quiet hamlet of Bethlehem, New Hampshire, into a thriving center of the hay fever tourist trade. During the 1860s, the widowed Helen Hunt moved after the first frost from her summer residence in the White Mountains to Newport, Rhode Island, where Colonel Thomas Wentworth Higginson, abolitionist, suffragist, and commander of a black soldier regiment in the Civil War, had welcomed Hunt into his literary circle and encouraged her aspiring career as a writer. Her first published prose, which appeared in the *New York Evening Post* in 1865, was an ode to the town of Bethlehem. Hay fever served as Hunt's muse.[15] Whether writing about the bewildering array of varied colors donned by the many trees and plants that performed in the autumn "miracle play" of Bethlehem or the power and sweetness of the Lord found atop Mount Washington, where only Mont Blanc to the east and Pike's Peak to the west rivaled its 6,285 foot elevation, Hunt earned the praise and admiration of America's leading transcendentalist poet, Ralph Waldo Emerson.[16]

Within a few years, however, Hunt's prose had helped transform her beloved rural haven of Bethlehem into a fashionable hay fever resort. In "A Protest against the Spread of Civilization," published in the *New York Evening Post* in 1867, Hunt wrote despairingly of how the quest for the

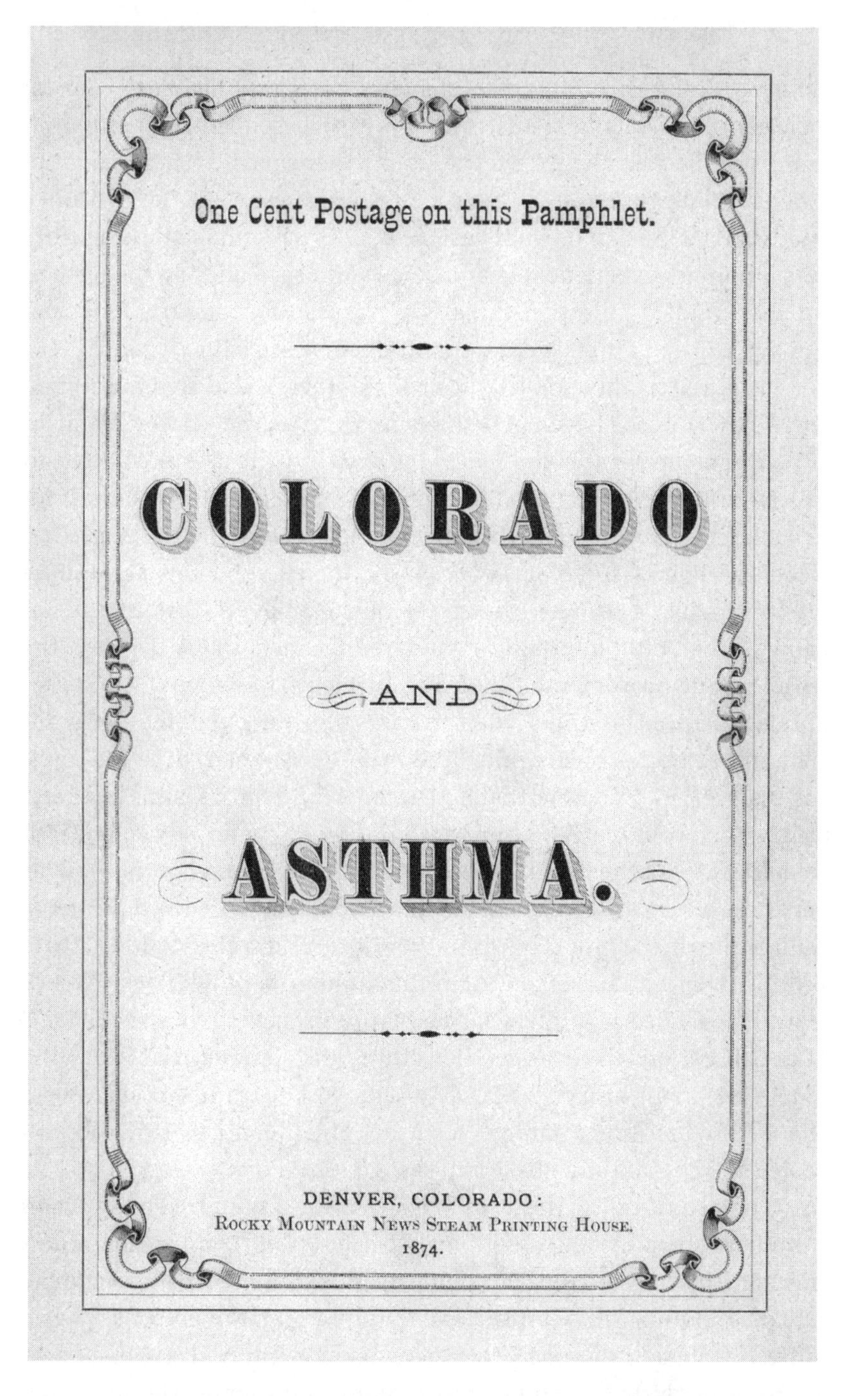

Figure 22. Brochures such as *Colorado and Asthma,* filled with testimonials about the therapeutic benefits of the Rocky Mountains, beckoned health seekers to the West. Courtesy of Wisconsin Historical Society, WHi-41446.

almighty dollar was driving Bethlehem's hotel proprietors to cater to increasing numbers of tourists in ways that despoiled the natural resources of the area and sullied the region's beneficial effects, as an antidote to civilization and its most notorious disease, hay fever. Hunt's essay marked the first in what became a series of periodic warnings that drew upon her experiences as an allergy sufferer to critique the ways in which unbridled industrial expansion was transforming the American landscape in ways that would jeopardize the health of its citizens.[17]

Dismayed by the rapidity with which progress and civilization had come to her bucolic haven of Bethlehem, Hunt journeyed west by rail in 1872, dispatched by the *New York Independent* to write a series of letters to eastern readers detailing the life and landscape along the western frontier. Like her dear friend Sarah Jane Lippincott, Hunt did not regale her readers with wild stories of the West—shoot-outs, robberies, and Indian massacres—that had been the staple of pulp novels. Instead she delighted them with entertaining tales and characters that depicted the "whole grand movement of the vast continent" westward.[18] Amid California's Sierras, Hunt rejoiced in the dry, pure air, exploring the "grandeur and solitude" of Yosemite—the majestic sequoias, dizzying trails, and magnificent waterfalls that had become John Muir's home and temple and that would catapult him to fame as America's leading defender of wilderness.[19] But the health Hunt found in the West did not follow with her as she returned east. A chronic sore throat, diagnosed as diphtheria, confined her to bed for months in the winter and spring of 1873. "Irresistibly impelled" to seek refuge in the sanitarium of the Rocky Mountains, Hunt asked Lippincott for assistance in making arrangements.[20] Ill health continued to plague her during the summer months as she sought respite in Amherst, Massachusetts, in a room free of dampness arranged by her friend Emily Dickinson. Finally, in late 1873, accompanied by her maid and physician, she left for Colorado.

When Hunt got off the train in Denver, she was greeted not by hues of brilliant sunshine and clear, blue skies, but by a "blank, bald, pitiless gray." She had crossed the continent, "ill, disheartened, to find a climate which would not kill." Instead, she found a "gray November sky" and suffered from a "rose cold." Hunt wondered if it would not be best to return to her New England circle of family and friends. Her physician advised her to try the recently established Fountain Colony, seventy-five miles south of Denver, also known as the town of Colorado Springs.[21]

Colorado Springs was only two years old when Hunt arrived via a spur of the Denver and Rio Grande Railway. In 1871, William Jackson Palmer, a Civil War hero and Union Pacific surveyor, along with his partner, Dr. William A. Bell, had amassed enough capital to build a narrow-gauge branch of the railroad from Denver to the newly created resort for wealthy invalids. Palmer's vision for a community of "people of culture and refinement," the "cream of eastern society," who were attracted to the health benefits of the region, allegedly came while sleeping under the stars, in the mountain air at the base of Pike's Peak.[22] Whether inspired by the smell of mountains or money, Palmer's grand scheme exceeded all expectations. Within a matter of six months, property values had tripled. When Hunt checked into the Colorado Springs Hotel, this temperance town boasted three thousand residents, three hundred homes, four churches, tree-lined streets, and land deeded for the establishment of Colorado College.

The disheartenment that gripped Hunt on her arrival in the Colorado Territory quickly gave way, dispelled by "delicious winter weather" that healed her body and soul.[23] Hunt showered effusive praise on the Rocky Mountain region, telling her friend Kate Field that here was "the divinest air I ever breathed."[24] She found in the region not only good health but also love. Among the boarders at the Colorado Springs Hotel was William Sharpless Jackson, a Pennsylvania Quaker who had taken up a position as secretary and treasurer of the Denver and Rio Grande Railway and oversaw the construction and management of the narrow-gauge line to Colorado Springs. Jackson spent the spring and summer of 1874 introducing his future wife to the region: they visited Sarah Jane Lippincott at her cottage in Manitou Springs, climbed in the mountains, and toured the mining district of Central City and Georgetown. Hunt, intending to stay on in a town that matched the elevation of Mount Washington, her favored New England retreat, ordered her trunks to be shipped west. By October 1874, "grilled with heat," she sought a change of place and left for a two-month vacation in Bethlehem.[25] But heart and health continued to beckon her west. She returned to Colorado Springs in December 1874 and married William Sharpless Jackson the following fall, taking up permanent residence in Colorado. Her seasonal sojourns in search of health had finally come to an end. Hunt Jackson quickly transformed the kitchen of their home into a living room so that she could revel in the view of the Rocky Mountains, "so spotless white, stately, and solemn, that if one believes there is a city of angels he must believe that these are the towers and gates thereof."[26] In Colorado, as in the White Mountains of New Hampshire, nature served as sanitarium, temple, and muse.

As Hunt Jackson assumed the life of a western woman, she looked with ambivalence upon the rapid development that was taking place before her very eyes. The influx of health seekers, prospectors, entrepreneurs, and investors to the Colorado Territory was in part the result of her own writings and the development of the railroads, but most important was the setting, one conducive to good health as well as bountiful in mineral riches. To many a nineteenth-century asthmatic who came to Colorado in search of relief, God had seemingly created, through the particular contours of soil, vegetation, and altitude, a place in the world where they could "breathe His air with comfort and ease."[27] Yet in the quest for material progress, Coloradoans ignored what to some was the most valuable natural treasure bequeathed by providence: health.

Smoke belching from mining regions of the Rockies signaled financial prosperity to civic boosters. But to health seekers like Hunt Jackson and Lippincott, the "yellow, suffocating smoke," which stood in stark contrast to the "buoyant and delicious" dry mountain air, served as a fearful reminder of where progress might lead.[28] Denuded hillsides stripped of timber and dirty mountain streams were already a common sight when Hunt Jackson settled in Colorado in the 1870s (Figure 23). By the early 1900s, only 20 percent of Colorado's original forests remained.[29] "Whatever gains in its material success," Hunt Jackson wrote of one small town's embrace of mining, "it will have lost something when the whistle of railroad trains and the noisy bustle of many people's living shall have driven off the antelope and the deer, which now come down to the river to drink."[30] Traveling through the dirty gulches, mines, and "toppling houses" of the mining district of Central City and Georgetown, Hunt Jackson predicted a future when the "mill-wheels will stand still; the mines will be empty; and pilgrims will seek the heights . . . not because they hold silver and gold, but because they are gracious and beautiful and health-giving."[31] When Colorado's economy faltered after silver prices plummeted in the early 1890s, Hunt Jackson reminded her fellow citizens once again that "the contagion of the haste to be rich is as deadly as the contagion of disease."[32] Keenly aware of the environmental and social impact of reckless mining and the precariousness of its future,

Figure 23. The environmental destruction wrought by mining in districts like Central City, Colorado, led some writers such as Helen Hunt Jackson to question whether Colorado wasn't sacrificing a valuable natural resource worth more than gold or silver: health. Courtesy of Denver Public Library, Western History Collection, X61823.

Hunt Jackson saw Colorado's destiny not in the overexploitation of its mineral resources but in the preservation and protection of its climate, where "asthma, throat diseases, and earlier stages of consumption" were, "almost without exception, cured by [the] dry and rarefied air."[33]

As an allergy sufferer, Hunt Jackson, like June Hewitt a century later, was attuned to aspects of both the built and natural environments—the quality of air, the presence of dust, smoke, and pollens—that evaded the senses of humans not so affected. She drew upon her disease experience to offer a cautionary tale about public health and public lands in the early development of the West at a time when the financial interests of mining seemed bright and when the search for health and wealth appeared compatible to railroad promoters and civic boosters. Smelter operations, like those of the Boston and Colorado Smelting Company, located in the suburbs of Denver, transformed precious ore in thirty large furnaces with smokestacks more than one hundred feet high into gold, silver, and copper worth $3 million annually. In the process, the company consumed more than one hundred tons of coal per day. As the smoke blackened the clear sky, others began to share in Hunt Jackson's concern. Perhaps this was a type of consumption that was not covered in the guidebooks and for which the climate of Colorado offered no antidote.[34]

Smelter plumes may have darkened Colorado's bright, sunny days, "famous all over the world." But just like the clouds of smoke billowing from the steam engines that carried visitors—six hundred a day in 1886—into Denver, they were signs of a city on the economic move.[35] Denver's business leaders continued to champion population and industrial expansion fueled by the extraction of natural resources from public lands instead of preservation of the region's health benefits. They did so even as the city found itself overcome by problems of smoke, overcrowding, and a lack of public parks, which sullied the very qualities of fresh air and expansive space that boosters had promoted in attracting the health tourist trade. The priorities became readily apparent when the Denver Chamber of Commerce hosted the Lands Convention in 1907 to mobilize western states in protest against President Theodore Roosevelt's conservation policies.[36]

Roosevelt's efforts to restrict grazing, timber, and mineral rights on western public lands were aimed at preserving the interests of the common man and the rights of future generations. But Roosevelt's push for conservation in the West was also shaped by the personal struggles he had endured as a severely asthmatic child and the many sojourns his family had undertaken to places where he might find relief. It was in the West, as a ranchman on the badlands of the Dakota Territory in the 1880s, that Roosevelt's recurring battle for health was finally won through the "superbly health-giving" benefits of the outdoor life and the emotional and physical freedom he experienced on the wide-open plains.[37] Roosevelt's love of the West and his passion for nature derived in part from his experience of illness. When Denver turned its back on Roosevelt's efforts to restrict the use of public lands, it also jeopardized the natural resources of air, climate, and sunshine that health seekers such as Roosevelt sought to preserve and on which the city was in large part built.

As civilization, with all its attendant problems, came to the Queen City of the Plains—Denver's population had reached 250,000 by 1920—the city of Tucson, to the southwest, positioned itself to become the new last resort for those suffering from asthma, tuberculosis, and arthritis. "So many cities built upon the resources of good climate have yielded to the pressure of commercialism," observed an ad for Tucson's Barfield Sanatorium in 1931, that "factories, irrigation districts or other business projects . . . have lessened the value of the city as a health center. Tucson has no such obstacles."[38]

Compared to Denver, Tucson by 1920 had in fact witnessed modest growth in the forty years since the tracks of the Southern Pacific Railroad had extended their reach in 1880 from San Francisco across the southern Arizona desert to what was then called the Old Pueblo, a small Anglo-Mexican community of seven thousand. Many looked to the arrival of the railroad as the end of the town's lawless reputation and Apache raids and the beginning of progress and prosperity. Although the major strikes of silver and copper in the Arizona Territory were almost one hundred miles to the southeast, near the towns of Tombstone and Bisbee, Tucson's nearby Silver Bell Mountains, Santa Ritas, Empire Mountains, and Sierritas yielded valuable ores of copper, lead, zinc, silver, and gold. For a time, the gleam of copper seemed to shine the path to Tucson's bright economic future, as the electrification of urban America created ever-increasing demand and escalating prices for the metal in the early twentieth century. Trains that turned Tucson into a transportation center of the Southwest also brought with them steam-powered pumps that greatly expanded the size of irrigated farms along the Santa Cruz River and transformed the desert into a center of cattle and cotton production. When the United States entered World War I, cattle, copper, and cotton were the major sources of Tucson's economic growth.[39]

Subject to the whims of weather and fluctuating cycles of supply and demand, neither agriculture nor mining proved capable of offering stable returns. After the war, huge oversupplies of copper led to significant corporate losses, massive unemployment, and labor strikes in the mining industry. The slump in copper production, coupled with depressed cotton prices and the emergence of Phoenix (surpassing Tucson) as the largest city in Arizona, left Tucson's business leaders wondering whether they had banked on the region's most profitable natural resources. By the 1920s, investment in climate topped the city's economic development plan. Civic boosters, hotel proprietors, sanatorium directors, prominent writers, scientists, and physicians worked to fashion Tucson's open space, unending sunshine, and dry desert air into a landscape of hope for those suffering from respiratory and rheumatic diseases. Cotton yields might falter, copper prices might drop, but the sunny climate was an enduring resource that Tucsonans could count on.[40]

The establishment of Tucson as the new climate capital of the United States and the building of a corresponding health and tourist industry depended greatly upon the advertising campaign launched by the Tucson Sunshine Climate Club in the 1920s. Formed in 1922 by prominent local businessmen, the club was an association governed by an unpaid board of directors; its express purpose was to make Tucson a permanent destination for health seekers and a winter vacation spot for tourists.[41] Since the 1890s, when market prices for silver had plummeted, Tucson's Board of Trade had promoted the area's "treasures of health" as unequaled "anywhere in North America." Unlike the climate of Colorado, the "pure, dry, invigorating air" of the desert was not subject to severe seasonal change, making it an ideal place for asthmatics and tubercular sufferers.[42] In 1890, John A. Black, Arizona's commissioner of immigration, cited meteorological statistics gathered by the U.S. Signal Service to assure the doubtful that "instead of being the fiery furnace which popular fancy has painted it," Arizona's climate was what the health seeker searched for in vain "on the wind-swept plateau of Colorado." In Arizona, Black exclaimed, "the blue skies of Italy, the balmy airs of the *Riviere,* the bright sunshine of Andalusia, and the bracing breezes of the American Alps" beckoned. It was a land "where health welcomes the afflicted, and where strength awaits the weak and the suffering."[43]

Such amateur promotional efforts, however, paled in comparison to the modern marketing techniques that the Tucson Sunshine Climate Club had at its disposal in 1922. With $37,000 raised for its first advertising campaign, the club hired H. K. McCann, an advertising giant that had pioneered motivational research and total marketing concepts. McCann launched a marketing and public relations blitz that reached 10 million Americans through large metropolitan newspapers, leading magazines such as the *Ladies' Home Journal,* and popular and professional medical journals such as *Hygeia* and the *Journal of the American Medical Association.* Tucson was a place where the "pale, inactive children" of eastern cities might grow into "robust vigorous youths" and where "hundreds of permanent, responsible residents [had] conquered pulmonary troubles, extreme nervous conditions, asthma, and other functional and physical disorders."[44] Through the ads and the distribution of brochures like "The Cure" and "Man Building in the Sunshine Climate," an estimated $2 million came to the local economy in the first two years of club activities. By the late 1930s, that figure had grown to more than $7 million in annual income from the health and tourist trade—second only to the revenues Tucson earned from Arizona's mining industry (Figure 24).[45]

Paid advertising that catered to those in search of health and relaxation was one sure-fire way to increase Tucson's population and economic growth. But nothing beat free promotional copy furnished by literary celebrities. In the novelist Harold Bell Wright, Tucson had such a person. What Helen Hunt Jackson had accomplished for the Rocky Mountain region forty years earlier Wright accomplished in the 1920s for Tucson—if not more.

Wright may have been shunned by the literary critics of his day, but his inspirational novels, centered on "right living," were immensely popular among ordinary folk. Factory workers, farmers, homemakers, and cowboys—people in "humble homes in out of the way American places"—found a compelling narrative in Wright's stories, with their disdain for the materialistic life fostered by urban civilization and their embrace of a simple, rural lifestyle and the appreciation of nature as guides to spiritual wealth. The nineteen books Wright published between 1903 and 1942 sold over 10 million copies; for almost twenty years his books outsold all others except the Bible, making him one of the most successful commercial authors in literary history. Thus, when Wright penned an article, "Why I Did Not Die," for *American Magazine* in 1924 extolling the health-restoring virtues of Tucson's desert environment, people took notice. Reprinted as a booklet by the Sunshine Climate Club and in installments in large metropolitan newspapers, the article reached an enormous number of readers, many of whom followed Wright's pilgrimage for health to the mecca of the desert sun.[46]

Wright began his career as an itinerant preacher of the Christian Church (Disciples of Christ) in the Midwest. Fragile health, however, prompted him to move to southern California upon the completion of his second novel, *The Shepherd of the Hills*. Published in 1907, the book catapulted the author to fame and enabled him to devote his full-time

Children of the Sun live here

Brown, sturdy, rosy-cheeked — growing into robust vigorous youths—Tucson's children flourish like flowers in the Sunshine-Climate. Outdoors all winter long—romping —hatless—they are the children of the sun.

For Your Child

Here is the place, Mothers, for your pale, inactive children. A season in Tucson—schools accept pupils any time— will bring big appetites and fill frail little bodies with glowing health and energy—the heritage of every normal child. Anxious mothers come each season and discover here restorative virtues quite beyond their anticipation.

Rents are Reasonable

You can come to Tucson and live here with only a moderate income. Rents are reasonable. Good hotels, churches, clubs, golf, horse-back riding, motoring, every city convenience.

Fares Reduced

Low rates are now effective via Southern Pacific, Rock Island and El Paso and Southwestern. Tucson is only fifty-three hours from Chicago, seventy-three from New York.

Fill out the coupon below and we will gladly send you "Man-Building in the Sunshine-Climate." It tells of the joy of living in Tucson.

Tucson Sunshine-Climate Club,
201 Old Pueblo Bldg., Tucson, Ariz.

Please send me your free booklet, "Man Building in the Sunshine-Climate."

Name ______________________

Address ______________________

Figure 24. The Tucson Sunshine Climate Club used the latest trends in advertising and public relations in the 1920s to turn Tucson into the climate and health capital of the United States.

energies to writing. For almost ten years Wright had suffered from recurring bouts of severe pneumonia. Over the years, physicians advised him that unless he sought a milder, warmer climate, he was unlikely to live much past his thirtieth birthday. When Wright received an invitation to be pastor of the First Christian Church in Redlands, California, he was quick to accept. And California's climate seemed to work miracles on Wright's frail body. Within a short time, he was riding horseback, rustling cows, and living the rugged life demanded by the ranch he had purchased east of El Centro in the Imperial Valley. Then, in 1915, while traveling on the road from El Centro, Wright and his beloved horse Mike were hit by an automobile. Mike suffered fatal injuries; Wright was more fortunate. Although he survived the accident, X-rays revealed that Wright had developed an active case of tuberculosis. Physicians advised Wright once again to seek a dry, mild climate and a life outdoors. "That spells Tucson, Arizona," Wright recounted in his 1924 *American Magazine* article.[47] He promptly headed to the Old Pueblo, set up camp at the foot of the Santa Catalina Mountains, dressed himself in white to benefit from the unadulterated sunlight, and began following a careful regimen of exercise, rest, and writing (Figure 25). Within five months, he found his health restored. Convinced of the region's therapeutic powers, Wright took up residence in Tucson three years later, building a desert home on 160 acres, five miles east of the city's edge.

Wright became Tucson's most outstanding personality in the 1920s, and his presence and stories of the Southwest helped transform what once appeared to be an inhospitable wasteland into a land of hope and renewal. He became a living testimonial to the therapeutic qualities of Arizona's sunshine and climate. In addition, the "grandeur and beauty of [the] vast desert" and the "wild, rugged mountains" that were the Catalinas he depicted in novels like *The Mine with the Iron Door* made the Sonoran Desert come alive for readers as an alluring, mysterious place.[48] In *The Mine with the Iron Door,* it is through Jimmy, a lunger living in a small white house on the mountainside above the Canyon of Gold in the Santa Catalinas, that the reader witnesses the "scarlet glory of the ocotillo" on the mountainsides, the "great fields of golden brittle-bush" on lower elevations, and, still further down, "the yuccas (our Lord's candles) in countless thousands, raising their stately shafts with eight-foot clusters of creamy-white bloom."[49] To Jimmy, and to Wright, this was

Figure 25. Harold Bell Wright's camp at the foot of the Santa Catalina Mountains. Courtesy of Arizona Historical Society/Tucson, AHS# 29004.

not a grim, dreadful, barren land but a place of healing. The desert's therapeutic benefits only added to its mystery and charm.

In the remote desert, Wright had found an antidote to the "driving tendencies" of his generation, the hurried, "terrific pace" of modern civilization that strained "human flesh and bone and blood." It was a natural place of healing that needed protection from the engine of progress driven by civilization. Like Helen Hunt Jackson and Teddy Roosevelt before him, Wright linked his interest in the conservation and preservation of Western lands to health. As one of several people, including fellow novelist Zane Grey, to lobby for the establishment of a national reserve in the Sonoran Desert region and as an activist in conservation organizations such as the Save-the-Redwoods League, Wright saw his dream partly fulfilled when President Herbert Hoover in 1933 created the Saguaro National Monument to the east of Tucson in the Rincon Mountains.[50]

Although Wright railed against material progress in his novels—it was the source of the "very diseases that rot the white man's bones, wither his flesh, dim his eyes, and turn his blood to water"—his fame helped to speed the pace of Tucson's growth.[51] At the same time, his affluence afforded him the luxury of escaping the encroaching ills of modern society that began again to plague him. In 1929, the *Tucson Citizen* credited Wright's "magnificent estate" on the outskirts of town as the beginning of a "transformation from untamed desert that could once be acquired for a song, to landscaped estates, handsome residences, and improved subdivisions."[52] Wright did not welcome the invasion of civilization into his idyllic retreat. Increasingly fed up with the clouds of dust caused by increased automobile traffic on Tucson's unpaved roads, urban sprawl that blocked some of his most beloved desert views, and a steady stream of tourists that infringed on his privacy, Wright left Tucson in 1936. He instead divided his time among Hawaii, Jamaica, and a remote estate he purchased north of San Diego. Wright may have had contempt for materialism, but his wealth offered him access to a geography of hope and healing that those less fortunate found more difficult to find.

In developing health industries on the Western frontier, both Denver and Tucson employed nature. "Nature alone cures," wrote one Tucson physician, "and where there is no nature there is no hope."[53] Health seekers looked to the unique physical geography and ecology of the Rocky Mountains or the Sonoran Desert for hope. It was on that hope that an economy catering to the needs of the chronically ill traded and thrived. Nature alone might cure, but physicians, sanatoria, hospitals, and rest homes were also ready to aid the afflicted—particularly paying patients—offering medical care that enhanced the healthful benefits of nature. By the early twentieth century, rail travel had put places like Denver and Tucson within reach of many invalids. Often, they could afford only a one-way ticket. But what need was there to return when they were headed to the promised land?

Although the mountains and desert did not discriminate between asthmatics and consumptives, town residents did. After the discovery of the tubercle bacillus by Robert Koch in 1882, awareness of the threat of contagion that tuberculosis posed to public health led to the physical and social isolation of tuberculosis sufferers from others afflicted with breathing difficulties. By the early 1900s, tuberculosis increasingly came to be seen as a disease of the poor, situated within the crowded tenements and squalor occupied by America's growing immigrant population. Not surprisingly, tuberculosis sufferers of little means who migrated west were often shunned by established residents. In Tucson, such patients found themselves restricted to any of several tent colonies, such as Tent City, which existed on the edge of town in the early 1900s. Adams Street Mission, not far from Tent City, was established in 1909 by the generosity of the Baptist preacher and printer Oliver E. Comstock. It was the only charitable "hospital" in Tucson. Even so, it wasn't much. The facility consisted of three tents staffed by volunteers and financed by whatever meager funds Comstock could personally give or raise from friends and local organizations. It was Wright who helped to greatly improve the care of Tucson's indigent invalid population. In the early 1920s, he staged a series of performances of his novels *The Shepherd of the Hills* and *Salt of the Earth.* Wright donated the proceeds to Comstock, who converted the tent hospital into a four-room masonry building, with a children's wing and an operating room run by Tucson's Organized Charities.[54]

Inadequate housing and medical care greeted the poor who came to Tucson in search of better health. But those who came with money found the sun's warmth matched by the hospitality and care that hotel proprietors, sanatorium directors, and hospital staff were more than willing to bestow upon paying patients. The successful promotion of Tucson as the climate and health capital of the United States precipitated a huge influx of seasonal visitors. In 1930, when Tucson's population was just over thirty thousand, the size of the town fluctuated by as many as ten thousand people between the winter high and the summer low.[55] Suddenly, like the flowering of the desert after a spring rain, an urban infrastructure that included twenty-one sanatoria, four hospitals, and four luxury hotels appeared in Tucson in the early 1930s. Some institutions, such as the Barfield Sanatorium, which occupied an entire block along the city's major boulevard, East Speedway, advertised care for those suffering from arthritis, asthma, hay fever, and other allergic diseases and refused those with advanced stages of tuberculosis. Others made no distinction between those afflicted with contagious and non-contagious diseases.[56]

While a number of these sanatoria were nothing more than large private homes with little or no specialized medical staff or equipment, others combined nature and technology to furnish elaborate medical care. The Desert Sanatorium and Institute of Research was Tucson's most famous state-of-the-art medical institution for the diagnosis and care of patients suffering from chronic arthritis and pulmonary disease. A 120-bed treatment and research facility, the "San" had a clinical and research staff of over thirty physicians and scientists; a national board of directors that included such medical luminaries as Dr. William Welch, director of the School of Hygiene and Public Health at Johns Hopkins University; and the latest in diagnostic and research equipment. Visited by eminent writers, European royalty, and movie actors (including Gary Cooper), the San was an exclusive resort, hospital, and scientific laboratory combined into one.

Built in 1926 four miles northeast of Tucson on 160 acres near the crumbling adobe ruins of the nineteenth-century military outpost Fort Lowell, the San took its inspiration from the region's natural and cultural heritage and united the enterprising vision of the physician Bernard L. Wyatt with the financial fortune of the legendary advertising businessman Alfred W. Erickson. In 1925, at a Christmas gathering at Erickson's New England game lodge, Wyatt, whose wife Minnie was a close friend of Erickson's wife, Anna Edith, convinced the Ericksons to contribute $25,000 toward the building of a medical establishment based on the therapeutic properties of sunlight. When Erickson's New York City agency merged in 1930 with H. K. McCann, the ad firm that had helped turn Tucson's climate and sunshine into valuable commodities, the Ericksons' investment interests in Tucson deepened. By the time of Arthur Erickson's death in 1936, the family had donated over $1.5 million toward maintaining and enhancing the San.

Wyatt was captured by the "strange potency of that Arizona mesa land, . . . where for years in increasing numbers . . . poured invalids and semi-invalids . . . to luxuriate in warm dry air . . . and in a baking sun whose blazing light has sent countless artists crazy in the endeavor to put on canvas the hard outlines and color of the southwest desert scenery." In Tucson, Wyatt saw great potential in the establishment of a treatment and research facility devoted to the scientific study of solar radiation and its physiological effects. Wyatt's model was Dr. Auguste Rollier's clinic in the Swiss Alpine village of Leysin; the clinic used high dosages of solar radiation, known as heliotherapy, to treat tuberculosis patients. Tucson was an ideal setting for heliotherapy in the United States. Not only did it offer unparalleled sunshine, but nearby institutions also offered a wealth of expertise. One such institution was the Desert Botanical Laboratory, created by the Carnegie Institution of Washington in 1903 and located outside of Tucson on Tumamoc Hill. There botanists such as Daniel T. MacDougal and Forrest Shreve were unraveling the unique ecology and physiological adaptations of plant life in the arid conditions of the desert. Such research proved vital in understanding the precise physiological effects of the desert's low relative humidity and high incidence of solar radiation on biological organisms, including humans.[57]

In the planning, building, and many expansions of the San, considerable care was taken to cultivate, not destroy, the desert ecology, which scientists and clinicians considered an important part of the healing services offered by the facility. Buildings were modeled after Hopi-style architecture because of its proven suitability to the "utilitarian and aesthetic requirements of the Desert region." The Native American theme was carried through in both the interior design, which featured Navajo rugs and mural reproductions of Zuni altar and ceremonial paintings, and the names of patient units, which included Pima, Papago, Navajo, and Apache—names meant to be reminders of the close relationship between the land and its former inhabitants. The dun colors and low, rambling buildings were designed to blend with the "cactus-clad plains" and "pine-crested" Santa Catalina Mountains to the north, so as not to "jar on the sensibilities of the invalid who seeks the solace of the Desert calm."[58] Harold Bell Wright, in the role of vice president of the board of directors, worked to landscape the property with his publisher's nephew, Rutger Porter, who was known to the sanatorium staff as the "Kactus Kid." Wright had been among an avant-garde group of wealthy and well-educated arists, naturalists, and writers who embraced Tucson's Spanish-Mexican heritage in architecture and landscape design. At his Tucson home, he preserved and planted native desert vegetation, adding over three hundred types of cactus. At the San, Wright expanded on his interests in native landscape design, creating at the time one of the largest gardens of rare cacti and native desert plants in the world. The Desert Sanatorium, in its construction and in its marketing to prospective patients throughout the United States, Mexico, England, and Europe, highlighted the restorative qualities found in the region's unique history, physical geography, and ecology.

Nature might furnish the raw materials of healing, but if unrefined, its potency was limited. For those with financial means, science and technology could harness nature and enhance its healing powers. Drawing upon the latest medical and scientific advances, the San sought to isolate, bring out, and refine the healing properties of sunlight and climate, just as miners extracted ore from the nearby mountain ranges and reduced it to precious metal. The main solarium, said to be the largest in the world built solely for sun therapy, used eight hundred square feet of a high-quartz-content glass specially manufactured in England. It reflected heat waves while allowing the penetration of ultraviolet light. Atop the building sat two copper-covered domes, mosque-like edifices to the sun. In one sat a radiometer, one of only two such instruments built by American astronomer Edison Pettit to measure fluctuations in ultraviolet radiation throughout the day. In the other sat a specialized astronomical mirror known as a Foucault siderostat, also built by Pettit, to catch and direct sunlight to a series of quartz prisms that allowed researchers to experiment on the physiological effects of different wavelengths of light on animals and plants. Although harnessing the sun for the clinical treatment of pulmonary tuberculosis patients proved ineffective—in 1928, the Desert San ceased to admit such patients—the results of heliotherapy in the treatment of patients with bronchial asthma, chronic bronchitis, and pleural tuberculosis seemed especially promising.[59]

The San represented a unique gathering of physicians, physicists, biochemists, and ecologists in therapeutic research and clinical treatment. As director of the Desert Botanical Laboratory and as a member of the San's board of directors, Daniel T. MacDougal promoted investigations into the relationships between the regional ecology and climate of the Sonoran Desert and its effects upon "health, human comfort, and general well-being."[60] Because nature was a valuable resource upon which the economic future of the region depended, it needed to be preserved. Health and conservation went hand in hand. Researchers at the San might find "the wealth hidden in the atmosphere of Tucson," but as Michael Pupin, the famous Columbia University physicist, solar radiation expert, and member of the San's research team, remarked, "They can do no good when nature is destroyed."[61] Pupin warned the people of Tucson in a 1929 article in the *Arizona Daily Star* not to do anything that "might destroy the wonderful qualities of the Tucson desert atmosphere." Excessive irrigation, factory smoke, and dust from increased automobile traffic were all potentially harmful to the region's fragile ecology and climate. While the details of the local ecology and climate remained largely a mystery to science, they were nevertheless vital, Pupin argued, to the continued development of Tucson as the most popular health resort in the United States.[62]

Research on desert ecology and human health, however, was short-lived. The bleak economic climate of the Depression forced the San to shut down its research institute in 1933. Wealthy patients, fewer in number after the stock market crash of 1929, continued to come, but the numbers arriving from abroad dwindled as the war in Europe began. Meanwhile, the influx into Tucson of more than ten thousand servicemen and their families at any of three flight-training centers operating there during the war years, in addition to the continued health migration, placed a strain on the city's hospitals. In 1943, Anna Erickson donated the San to the citizens of Tucson. With the aid of experienced fund-raisers, such as the birth control activist Margaret Sanger, Tucsonans raised an additional $250,000 to convert the San into a nonprofit community hospital, the Tucson Medical Center.[63]

The conversion of the San into the Tucson Medical Center was a portent of where the new treasures in Tucson's expanding health economy would be found. Arizona sunshine nourished the bodies of patients who came to the San, hopeful that the desert's curative powers might give relief from chronic illness. At the same time, the patients' diseased bodies became resources for institutions like the San and came to support the health tourist trade in much the same way that Arizona's climate and sunshine first had. Within the walls of sanatoria and hospitals, a new space of hope was gradually taking shape, built around the promise of biomedical research. As the health migration to places like Tucson and Denver altered the regional ecology and its perceived therapeutic benefits, those who sought relief in the desert sun or mountain air increasingly found hope in the wonders of biomedicine rather than the healing powers of nature.

At first, the dwindling supply of diseased bodies threatened the health economy of Denver and Tucson far more than landscape change, despite the periodic warnings of health seekers and scientists. In the early twentieth century, Denver had twice the national average of physicians per capita and was home to numerous sanatoria and hospitals, including the National Jewish Hospital, the Jewish Consumptive Relief Society, and Fitzsimons U.S. Army Hospital. Like Tucson, Denver had a thriving economy that depended upon a continual traffic of health migrants, mostly those suffering from pulmonary troubles. Tuberculosis sufferers were the most visible among the first wave of invalid immigrants who flocked to Denver and Tucson. Asthmatics came as well, but since their disease posed no contagious threat, their presence was less visible and their numbers less certain. By the 1930s, however, Denver's and Tucson's tubercular tourist trade had begun to decline. One reason for the decline was a widespread public health campaign by the National Tuberculosis Association and state health departments that actively discouraged tubercular patients from seeking a change of climate as a remedy for their illness. Such campaigns were an outgrowth of the closed institutional model of sanatorium care that had developed for TB patients. With its regimen of rest, nutritious diet, and fresh air, along with surgical intervention that collapsed an infected lung to facilitate healing, institutionalized medical care became far more important in TB treatment than physical geography.[64] By World War II, empty beds bespoke a void in the institutional marketplace around which the health economies of Denver and Tucson were built. The disease that would fill that economic void was asthma. But first the market had to be made.

Unlike tuberculosis and polio, asthma was not a disease in the national spotlight at the time of World War II. Two institutions—Tucson's National Foundation for Asthmatic Children (NFAC) and Denver's Jewish National Home for Asthmatic Children—helped change that. The visibility of asthma, which in the early 1950s afflicted roughly 2.5 million children in America, expanded with aggressive national marketing campaigns that drew public attention to the plight of the asthmatic child and to Tucson and Denver as national centers for asthma treatment and research.[65]

Established in 1949, the NFAC was the first charitable disease-based organization to launch a national campaign on behalf of the asthmatic child. A decade before its founding, Raphael Brandes and his wife brought their experience in social work and teaching, gained in New York City, to Tucson and founded a private boarding school for asthmatic children. As the head of Tucson's Jewish Community Council, Brandes was well aware that of those who moved to Tucson for health reasons and sought aid from the Tucson Jewish Relief Society, the majority were asthmatic. Brandes wanted to provide an opportunity for less fortunate families with asthmatic children who could afford neither the financial burden of moving to the Southwest nor the tuition of a private boarding school. In 1949, he met with a group of parents who were impressed by the "miracle" the Arizona desert and the Brandes school had worked on their children's sickly bodies. The result was a nonsectarian, philanthropic organization, the NFAC. In a decision that they hoped would make Tucson "as well known eventually as Warm Springs, Georgia, is . . . for the treatment of polio victims," the foundation directors chose the Old Pueblo as the site for a national center for asthma education, research, and treatment. The center also included a sixty-bed residential rehabilitation facility for low-income asthmatic children who had failed to respond to conventional medical therapy (Figure 26).[66]

Denver's Jewish National Home for Asthmatic Children was a 112-bed residential treatment facility with 62 staff members, including doctors, nurses, house parents, social workers, and psychiatrists. In the 1950s, it embarked on a national fund-raising drive similar to Tucson's to turn the Mile-High City into an institutional center for the research and treatment of asthma. The home was originally established in 1907 as the National Home for Jewish Children, a shelter for orphans or dependents of poor East European immigrants who had come to Denver in search of a cure for tuberculosis. By the 1940s, the National Home increasingly found itself providing care for asthmatic children, and in 1953 it officially changed its name to the Jewish National Home for Asthmatic Children to reflect the demographic shift in its patient base. In 1956, the home would expand and incorporate a new facility, the Children's Asthma Research Institute and Hospital (CARIH) in "the field of asthma and related diseases."[67]

In their efforts to publicize asthma and create a market for asthma treatment and research, both the NFAC and the Jewish National Home for Asthmatic Children looked to the National Foundation for Infantile Paralysis (NFIP) as a model. The NFIP knew about public relations and advertising campaigns to publicize a disease, and it knew how to open the pocketbooks of Americans. Established in 1938 by President Franklin Delano Roosevelt, the NFIP, through its close ties to radio and Hollywood, grew into a voluntary health organization that raised $20 million annually. Comedian Eddie Cantor, playing off the popular newsreel feature *The March of Time,* coined the phrase "the March of Dimes" during a Hollywood fund-raiser for the NFIP, urging radio listeners to send their dimes to help those with polio.[68] By the 1940s, the NFIP March of Dimes

Figure 26. Asthma convalescent homes, like the National Foundation for Asthmatic Children in Tucson pictured here, were often located in places where the climate had attracted those suffering from respiratory illnesses. Courtesy of University of Arizona Library Special Collections.

campaign had grown into an annual national media event that was orchestrated by well-honed public relations, radio, TV, and motion picture departments. Celebrities were featured in spots on radio and television and appeared in films. NFIP's local chapters were provided with slick campaign guides with information on the latest in organizational techniques and publicity strategies, including sample newspaper stories, departmental window displays, billboards, and construction plans for wishing wells and "Mile O' Dimes" stands.

The NFAC and the Jewish National Home followed suit. The NFAC looked to celebrities such as Jack Benny, Eddie Cantor, and Edward G. Robinson; politicians such as Senator Barry Goldwater and Arizona governor Howard Pyle; prominent businessmen like publisher William R. Mathews; and leading allergists such as Samuel Feinberg and Murray Peshkin to serve as members of its campaign committee. Radio and television spots by stars such as Steve Allen, public appearances by popular singers such as the Lennon Sisters, and direct-mail campaigns that solicited donations from over half a million contributors became a regular feature of the NFAC's annual fund-raising efforts.[69] Central parts of the Jewish National Home's campaign in the early 1950s were national direct-mail campaigns, auxiliary conventions, and a media blitz that

Collier's, April 16, 1954

They've Got Asthma on the Run

Classed as incurables, the children arrive at the home wheezing pitifully with every breath. Yet a year later they're healthy and active. What's the secret?

Figure 27. Groups like the Jewish National Home for Asthmatic Children looked to the success of the March of Dimes in using children to garner public attention and financial resources for research into the causes and treatment of asthma after World War II. *Collier's*, 16 April 1954.

included features on the home in national magazines such as *Collier's*, *Newsweek*, and *Parade* (Figure 27).[70] Although neither the NFAC nor the Jewish National Home could boast anywhere near the success of the March of Dimes' multimillion-dollar fund-raising campaign—both averaged annual charitable contributions on the order of $500,000—these efforts brought national attention to asthma and to Denver and Tucson as places where American children suffering from the disease might find a chance to "breathe again, to play again—to LIVE AGAIN."[71]

The focus on childhood asthma in these national campaigns was no coincidence. The NFIP owed much of its success to the emotional appeal of a child on crutches and in braces. A new March of Dimes poster child was selected each year. Although asthma's effects were far less visible than those of polio, both the NFAC and the Jewish National Home sought to portray asthma as a childhood crippler, a disease that claimed twice as many lives as that of polio and led to untold suffering on the part of the children and their families. Dr. Murray Peshkin, chief of Mount Sinai Hospital's Children's Allergy Clinic in New York City, asked the following question: What was to be done for "this pulmonary cripple who has been swept aside by a fast moving world into the gutter of despair, gasping in the shadow of hopelessness, feebly holding on and with tired lusterless, fading eyes, scanning and search[ing] the distant horizon for the sunshine of hope?" His answer was to send such a person to Denver or Tucson, where hope could be found not only in the physical landscape but, more important, in the institutional settings created by the National Jewish Home and the NFAC.[72]

Of the approximately one dozen convalescent homes for asthmatic children established between the late 1930s and early 1960s, at least one-third were institutions founded in places where the physical environment had long attracted those suffering from pulmonary illnesses.[73] Indeed, both Tucson's NFAC and Denver's Jewish National Home stressed the "warm, dry, sun-filled air" of the desert or the "dry mountain air" of the Rockies as reasons for their location and rehabilitative success. But gradually a therapy took hold that was originally proposed by Peshkin in 1930; in it the emotional environment of the asthmatic child drew far more attention in management and treatment than the physical environment. "Sensitizing substances"—pollens, dust, and other known allergens—were, Peshkin argued, "merely exciting factors, not the basic cause of the symptoms" found in the asthmatic child. Peshkin believed allergy to be the result of a "physiochemical disturbance [that] manifests itself through the nervous system." For children who did not respond to immunotherapy or adrenalin injections within their home environments, Peshkin sought to restore the "physiochemical balance" by separating the children from their parents; he called the therapy "parentectomy."[74] He placed the children, for six months or longer, in an institutionalized setting prepared to accommodate the needs of allergic patients. Peshkin recounted in an anecdote how he advised one family to send their son to a "high and dry altitude." To his chagrin, the father packed up his entire family and moved west. "This was a tragic blunder," Peshkin exclaimed, "because the boy required not only a change of climate and environment, but also medical supervision and separation from his parents, in other words, 'parentectomy!' "[75]

Children with intractable asthma seemed to respond favorably to this environmental change. In many cases, their clinical symptoms disappeared without medication. However, a decade would pass before homes were established that could follow Peshkin's radical course of therapy en masse. The NFAC and the Jewish National Home—with their close ties to the charitable organizations, social work agencies, and hospitals with which Peshkin associated in serving the needs of New York City's Jewish population—were among the first.

By the 1950s, as the psychiatric community came to endorse medical theories like Peshkin's, one could pick up any popular health magazine and read that "deep seated emotional conflicts between the child and [his or her] parents" or an "over-protective mother" played an instrumental role in the development of childhood asthma. Peshkin himself described asthmatic children as those "who have become psychologically allergic to their environments." No wonder then that the number of psychiatric and social work staff at both the NFAC and the Jewish National Home in the early 1950s far exceeded the number of attending physicians.[76]

Ecology and climate may have helped drive families with asthmatic children west, but the children who came seeking relief were the most profitable resources mined by Denver and Tucson, and they fueled two rapidly expanding health care economies. Disease lobbies, eminent scientists, television and movie celebrities, and politicians helped to push forward unprecedented outlays by the federal government and the pharmaceutical industry in biomedical research after World War II. In 1947, the budget of the National Institutes of Health was a meager $8 million. Two decades later, President Lyndon B. Johnson proclaimed the first week of May 1967 National CARIH Asthma Week to highlight medical research, "valiantly seeking the cause and cure of asthma"; by then total NIH expenditures surpassed $1 billion, of which $90 million went to the National Institute for Allergy and Infectious Diseases.[77] In isolating emotions from the complex of environmental factors that contributed to the severity of asthma and in adopting a rehabilitation program focused on a closed, institutional model of medical care, clinical allergists banked on biomedical technology to engineer a breathing space once sought in the healing places of nature. It was a space located not outside the body but within it. In the interrelationships of the immune, nervous, and respiratory systems and the bio-physico-chemical processes through

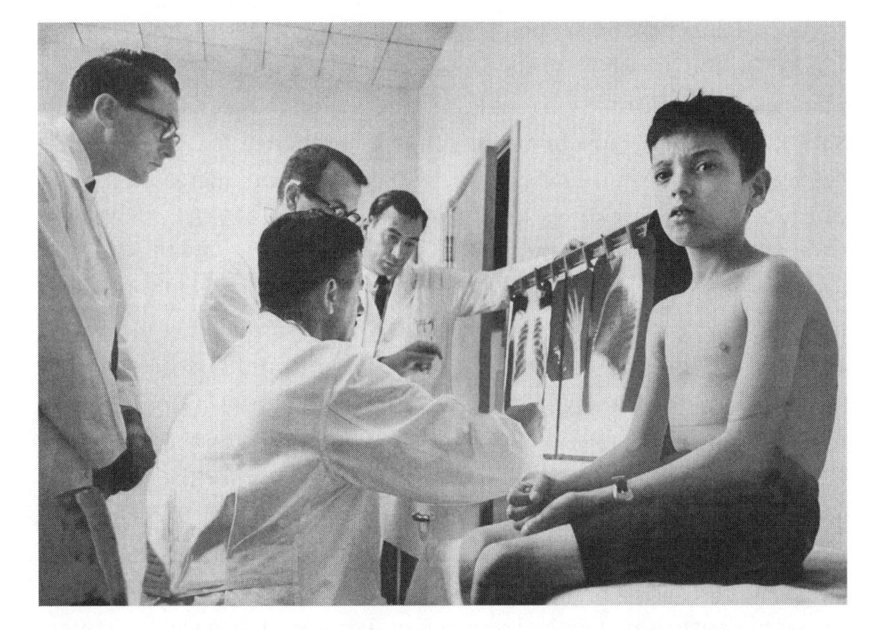

Figure 28. Children's bodies offered a new resource, and biomedical research offered a new source of hope, for the health economies of Denver and Tucson as the therapeutic qualities of nature waned. Copyright The Gordon Parks Foundation.

which oxygen and drugs flowed, a lucrative transnational market in pharmaceutical research and treatment of allergic disease would be built (Figure 28).

Places like the Jewish National Home in Denver presented an "ideal laboratory" for biomedical research on children's asthma.[78] Few other U.S. hospitals or convalescent homes housed such a large number of asthmatic children who were under daily medical supervision and stayed an average of two years. To clinical researchers, the children proved model research subjects. Not only were the children not afraid, having been poked and prodded by medical experts all their lives, but the social service staff had done a remarkable job of "conditioning these children for the slight, but very real, pain, which all patients have to experience in being scratch tested, having blood taken, [and receiving] the continuous injections which have to be given over a long period of time for inhalants to do any good."[79] Pharmaceutical companies quickly saw the research opportunities presented by such model subjects as they sought to test the first generation of antihistamines and corticosteroids, such as hydrocortisone, that were being developed for allergy sufferers. During the 1950s, the Jewish National Home had agreements with such major pharmaceutical firms as Charles Pfizer, the National Drug Company, and Smith, Kline, and French to test drugs in exchange for free pharmaceuticals and research grants.[80] Such research grants increased significantly when the Jewish National Home established CARIH. Aided by a $100,000 grant from the U.S. Public Health Service, CARIH opened its doors in 1959 and within a decade had grown into a world-renowned residential treatment and research center with an annual net income in excess of $1 million, mostly from federal grants and charitable donations. When CARIH merged into Denver's National Jewish Hospital (unrelated to the Jewish National Home) in 1978, it secured the Mile-High City's reputation as the premiere institutional center for the treatment and study of asthma and allergic disease.[81]

The NFAC was less successful in exploiting the resources offered by the children housed within its walls. Unlike CARIH, which considered the rapid expansion of federal research dollars and pharmaceutical interest in allergic diseases the key to its success, the NFAC continued to bet its economic future on Tucson's healthful climate. William R. Mathews, editor and publisher of the *Arizona Daily Star* and honorary co-chairman of the NFAC, made a remark in 1953 reminiscent of town boosters in the 1920s. He heralded Tucson's "contribution toward the health of the nation" as its "greatest industry" and praised the way the NFAC was "capitalizing on Tucson's health and climate."[82] Although the NFAC school had a medical director in place, it had neither the active clinical research program nor the medical facilities CARIH had by the 1960s. As long as the perceived therapeutic powers rested in place, what need was there for the physician? In 1964, the school moved, in part because increased traffic and residential developments near its downtown location necessitated the search for an "air-clear, dust-free environment." Only after opening a new million-dollar complex seven miles west of the city at the foothills of the Tucson Mountains did the NFAC begin to invest in clinical research—and it did so only because a $1 million bequest from Dr. Grafton Tyler Brown mandated that the funds be used for clinical research and treatment of adult allergic diseases.[83] In 1965, the NFAC set aside $250,000 of the Brown bequest to help fund the University of Arizona's medical program in the field of basic sciences and research into allergic diseases of adults. It used the rest to establish an adult outpatient clinic for asthma and allergy sufferers.[84] While the University of Arizona's medical school did eventually capitalize on the bodies of allergy and asthma sufferers in its midst, the NFAC never successfully made the transition to biomedical research. In 1969, plagued by a continual lack of funds and periodic scandals regarding the investment return on its fund-raising drives, the school of the NFAC closed its doors. A special report on the closing in the *Arizona Daily Star,* seemingly oblivious to Tucson's growing environmental problems, ended by stating that of the children who "have to leave Arizona . . . their asthma stands a good chance of becoming worse."[85]

Allergic bodies not only opened up new vistas on the frontiers of biomedicine, but they also brought into sharp relief changing land-use patterns that destroyed the healing properties of place and reinforced the need for a new space of hope within the body's immunological landscape. Deteriorating air quality was one perceptible change; it manifested as a cloud of photochemical smog that stung eyes and aggravated lung inflammation as asthmatics struggled to breathe. In Denver, as in other American cities, many of the air pollution problems resulting from photochemical reactions in the atmosphere were appearing in the 1950s.[86] The pollution not only reduced visibility but also posed significant health threats to those suffering from cardiac and respiratory diseases. In 1956—the same year CARIH incorporated to offer new hope, independent of Denver's once salubrious environment, to those suffering from chronic respiratory disease—citizens and public officials pushed for the first of a series of extensive studies of Denver's air quality. By the mid-1960s, a combination of factors, including the city's rapid growth; residents' heavy reliance upon automobile transportation; and Denver's geography, which allowed frequent temperature inversions, resulted in carbon monoxide and photochemical ozone problems equal to or greater than those in much larger metropolitan areas like Los Angeles or New York City. Denver became as famous for its "brown cloud" as it once had been for its "clear skies" and "fresh mountain air."[87]

Such environmental changes, however, also provided new opportunities. Denver's semi-arid climate and constant sunshine, once hailed for their therapeutic benefits, accelerated the photochemical transformation of pollution gases. The urban ecology of the city, which included a larger than usual population of respiratory illness sufferers, made Denver a natural site for research into the health effects of air pollution. In the early twenty-first century, at the National Research Center for Environmental Lung Disease, based at National Jewish Hospital, clinical researchers employ Denver's unique climate and geography to investigate the health effects of air pollution on patients with preexisting pulmonary disease and to determine the effects of air pollution on the respiratory system.

In Tucson's pro-growth postwar environment, business and civic leaders denied that air pollution problems had come to affect the quality of life in their cherished Old Pueblo. But Tucson was no longer the provincial town of forty thousand people that it had been before World War II. Welcoming and actively promoting the sunbelt migration, Tucson's Chamber of Commerce saw nothing but economic prosperity in the city's fivefold increase in population and eightfold increase in size during the 1950s.[88] In 1960, over two hundred thousand people lived within Tucson's city limits, an area of more than seventy square miles. The mining and defense industries attracted many. But tourism, valued at $900 million in 1959, continued as a mainstay of Tucson's economy.[89] And health seekers remained a vital part of the tourist trade. Roughly 30 percent of the population that migrated to Tucson during the two decades after World War II came in search of health.[90] When a member of the newly formed western section of the International Air Pollution Control Association suggested in 1957 that smog had come to this health oasis, Charles Arnold, manager of the Tucson Sunshine Climate Club, was perhaps blinded by the memory of a once bright and clear climate. He retorted: "Whoever said we had smog does not know Tucson." Air pollution was inconceivable, the *Arizona Daily Star* reported, to "the men who beat the publicity drums promoting the community as the sunshine and clean air Mecca of the world." But it was not so to those whose bodies were testimony to the ecological transformations of the desert wrought by sunbelt sprawl.[91]

Joseph Wood Krutch was one of many asthmatics who came to Tucson in the boom years of the 1950s, only to become dismayed at how quickly the mantra of progress and profit had altered the desert landscape and its therapeutic value. Columbia professor, *Nation* drama critic, biographer, and magazine writer, Krutch made a literary splash with the publication of his 1929 book, *The Modern Temper.* In what one reviewer praised as "one of the crucial documents of his generation," Krutch captured the moral dilemma of the modern age, where radical skepticism ruled over moral certainty, and science had thrust humanity into a godless universe, leaving an abyss of disillusionment and despair. This dilemma was unresolvable, Krutch argued, and it offered no escape.[92] Twenty years later, however, Krutch did escape. Worsening health, brought on by recurrent bouts of asthma, prompted Krutch to follow his allergist's advice and abandon the life of a literary highbrow in New York and move to Arizona. In 1950, he took a leave from Columbia and headed for the Sonoran Desert in search of regeneration. Never would he return to New York City, a place he came to describe as a "desert far more absolute than any I have ever seen in the Southwest." The sparseness of life, the vast open space, and the warm dry air brought not only physical renewal—Krutch claimed his health to be the best in years—but also a deepening appreciation for the ecology of the desert and the moral and spiritual meanings it conveyed. In books such as *The Desert Year* and *The Voice of the Desert,* Krutch portrayed the beauty of this arid environment. It was an aesthetic founded upon scarcity that Krutch gradually developed into a social critique of American affluence and abundance. "Much can be lacking in the midst of plenty," wrote Krutch. "On the other hand, where some things are scarce others, no less desirable, may abound." Through his nature writings and involvement with the Arizona–Sonora Desert Museum, Krutch became one of the most celebrated interpreters and champions of desert life.[93]

Like many a health seeker before him, Krutch came to his passion for the desert and a conservation ethic through his illness and the landscape in which he found relief. But it was a landscape being rapidly transformed by housing and commercial development, which altered the regional ecology and its health benefits. By the early 1960s, Krutch came to question in books and articles the pro-growth mentality embraced by Tucsonans. Cognizant of the privileged position he had in living in the desert (rather than having to make a living out of it), Krutch saw a tragedy unfolding as the southwestern environment became "increasingly used and exploited in ways peculiarly unsuited to, and sometimes completely destructive of, its unique character." In fewer than a dozen years, Tucson was "no longer recognizable as the town" in which he had come to live. "Large areas of what were then open deserts," Krutch lamented, "have been stripped of their natural growth, covered with hastily built houses crowded upon one another and, at best, now surrounded by struggling grass plots which not only require the expenditure of precious water but are aesthetically unsatisfying because they simply do not fit the landscape." The Catalina Mountains were "blurred by haze, and from the summit one looks down upon a pall of smoke and dust which lies like an ugly lake over the town. . . . Newcomers are quite right to ask me," Krutch opined, "where are the clean pure air and the sparkling nights which I described so enthusiastically in my first book about the desert. 'Progress' has taken away from me both my invigorating air and my brilliant stars."[94]

Progress was indeed changing the special character of the desert in ways that made Tucson's allergy and asthma population especially vulnerable. This was particularly true for those in lower socioeconomic groups, who moved to Tucson in increasing numbers as health care costs escalated. They came, believing as the Chamber of Commerce brochures espoused, that "Tucson's healthful climate" would miraculously relieve their "pain and suffering." They found instead pollen, mold, and smog.[95]

The obscured views of the Catalinas were no figment of Krutch's imagination. On a winter day in 1961, a charcoal-tinted sky hung over the Santa Cruz Valley. On a clear day, Tucsonans could see as far as one hundred miles into the distance, but on that January day visibility was restricted to less than three miles. Within a few weeks, experts testified before the Arizona House Planning and Development Committee that smog had indeed come to the town advertised as the "health capital of the world."[96] Although observed oxidant concentrations never reached the alarmingly high levels found in cities like Los Angeles, the fact that there were days when concentrations exceeded those often associated with eye irritation was cause for concern. Quentin Mees, a civil engineer at the University of Arizona, first formally alerted the city to the potential smog problem, which he said was due to the unique meteorological and topographical features of the Santa Cruz Valley and its expanding population. He remarked that "for a proven health center and a growing industrial center such as Tucson" to allow the problem of air pollution, which was both a "public health and economic problem," to "increase without surveillance should be unthinkable."[97]

Four years earlier, Tucson had buried its head in the sand at the first suggestion that its air quality might be a growing issue. Once they understood the potential economic impact, however, Tucson officials helped fund a four-year monitoring study undertaken by Mees. The study pinpointed the automobile as the single major source of Tucson's air pollution problem and recommended a community-wide approach of "continued caution."[98] But Tucson could still boast moderately good air quality compared to cities like Denver. Despite Mees's warning, development continued unabated. By the late 1970s, when the incidence of asthma in Tucson was twice that of the national average, the city found itself consistently in violation of National Ambient Air Quality Standards; the violations jeopardized the health of its residents suffering from respiratory disease.[99]

Nitrogen dioxide, ozone, and particulates generated from automobiles were not the only risk to Tucson's allergic citizens. Another was the pollen emanating from grass lawns and trees lining city streets. Much of the native vegetation in the desert and arid mountains when health seekers first came to Tucson consisted of creosote bush, numerous cacti, and other xerophytic flora (adapted for life and development under conditions of limited water supply) that relied upon animal and insect pollination for survival. Only a few wind-pollinated species—in particular the false ragweeds and saltbushes—were allergenic. It was Tucson's unique desert ecology that bolstered the city's reputation as a haven for allergy sufferers.[100]

Rapid urbanization changed everything. Ever since Harold Bell Wright had ushered in subdivision development of the desert in the 1920s, desert flora had begun to change. Few Tucsonans found in native desert vegetation the same aesthetic and spiritual values that Wright saw when he landscaped his home and the Desert Sanatorium in keeping with the ecology of desert plant life. In the housing developments that began in the 1920s and exploded after World War II, it was the "civilized" look of eastern cities, from which many residents came, that was favored. Front lawns were planted with Bermuda grass, and streets were lined with exotic evergreens, eucalyptus, African sumac, and silk oak. These plantings increasingly altered the desert ecology and reduced what were its healing properties.[101] Bermuda grass, aided by frequent watering, readily adapted to the desert soil and changed not only the look of the desert but the content of its atmosphere. Bermuda grass flourished in town lots and proliferated in tract home subdivisions and mobile home and retirement parks on the city's edge. It also made its way along watercourses into higher and cooler climatic zones. It was the first of the most potent allergens to be introduced into the desert landscape, and Krutch found it aesthetically repugnant.[102]

Eastern tastes for shade created a nursery business in exotic ornamental trees. Two species in particular wreaked havoc on the desert's therapeutic properties: mulberry and olive. Mulberry, which originated in East Asia, is a fast-growing tree quick to produce shade. It seemed ideally suited to the needs of the trailer courts and mobile home parks that sprung up after World War II. Olive trees, introduced in the early part of the twentieth century, became a favored choice among builders of shopping malls and apartment units in the postwar development boom. Both tree species are highly allergenic. The increasing use of these and other ornamental trees also created an abundant substrate for growth of the mold *Alternaria,* by now recognized as a major allergen associated with childhood asthma.[103]

Tucson's ornamental tree population reached maturity in the 1970s. During the same decade, grass pollen and mold spores increased. Native weed species, like tumbleweed and desert ragweed, which thrive in disturbed soils, likewise spread. Thus the former ecological haven had become an ecological hell. In a little more than twenty years, the atmospheric pollen load of allergenic plant species in Tucson had increased tenfold. Not only was the incidence of asthma now twice the national average, but also the incidence of hay fever was six to nine times greater.[104] The migration of allergy and asthma sufferers to the area alone could not account for such disparities; environmental changes played a significant role. Allergic citizens, dismayed by the ecological catastrophe inflicted upon their health, united. Led by retired postal worker Herman Berlowe, they lobbied the Tucson City Council in October 1975 to enact a ban on the sale of nonindigenous pollen-producing trees within city limits.[105] In the more exclusive subdivisions in the foothills, where trends in native landscape design were more prevalent than in middle- and lower-class neighborhoods, deed restrictions against the planting of Bermuda grass already existed to cater to wealthy homeowners seeking a "pollen-free, dust-free" zone.[106] Nine years later, against the objection of some nursery owners and developers, the city adopted a ban on the sale and planting of mulberry and olive trees and an ordinance requiring the frequent mowing or removal of Bermuda grass lawns. It was but a sad reminder of how, despite the repeated warnings of health seekers and scientists, the valuable ecology and health of a region had been squandered, leaving no safe place to turn.

Over the course of a century, the hope once found in the western landscape had been washed away, carried by the flood of people, plants, industries, and transportation that came with progress. Seeking relief from allergic disease, America's mobile suffering citizens had first turned to nature in their quest for an antidote against this malady of civilization. It was upon nature that the initial health economies of Denver and Tucson were built. But as the therapeutic value of these health-giving environments waned, hay fever sufferers and asthmatics found themselves out of place in the locales to which they had come as a last resort. The railroad had once carried them to the promised land. Now medical technologies paved the road to a new Shangri-La—with antihistamines, corticosteroids, and institutionalized care. The diseased bodies upon which these medical technologies worked had become a new natural resource for cities that still sought to attract the health tourist trade. As the twenty-first century dawned, Denver and Tucson found themselves home to world-renowned institutions—the National Jewish Hospital and the Arizona Medical Sciences Center—for research and treatment into respiratory and allergic disease.

But the geographies of hope to be found in either nature or technology have been equally blind to the ecology of justice. The wealthy first reaped the rewards of Denver's and Tucson's salubrious environments. They were the ones who could afford to escape the polluted, crowded, industrial atmosphere of eastern cities, just as they would later be the ones with easiest access to the latest biomedical advances in the battle against allergy and asthma. For the poor asthmatic child who lacked the good fortune to be one of the seventy-five out of one thousand children admitted into CARIH or similar institutions—where the annual cost for treatment in the late 1960s averaged $6,000 per child—the only hope was the arid Southwest. Contrary to evidence in the air that Arizona asthmatics breathed, Tucson never stopped selling itself as "the health capital of the world." "Moved down to Tucson Arizona in hopes of clear sky, warm temperatures and NO ASTHMA!" exclaimed a former CARIH patient in 1980 after losing access to quality medical care and again being beset by severe asthma.[107] A Tucson Web site in the spring of 2006 reminds us, "The City's dry desert air and winter sunshine make it a popular health and winter resort."[108] It is upon hope and past historical landscapes that today's lucrative allergy industry is built.

CHAPTER

4

Choking Cities

Man! How many times have I stood on the rooftop of my broken-down building at night and watched the bulb-lit world below. Like somehow it's different at night, this my Harlem. There ain't no bright sunlight to reveal the stark naked truth of garbage-lepered streets. Gone is the drabness and hurt, covered by a friendly night.

—Piri Thomas, 1967

On 7 July 1961, a forty-six-pound, malnourished twelve-year-old boy from a Rio de Janeiro *favela*—or slum—stepped off a plane in Denver, hoping to escape the shackles of poverty and longing for a life free from the struggle to breathe. It was just one week after photojournalist Gordon Parks's exposé of life among Latin America's urban poor had appeared in *Life* magazine. Flavio da Silva had been living with his family in Catacumba, one of the many squatter settlements of Rio. The family had fled the rural poverty of northeastern Brazil in search of a better life but had found instead—like many of the *favelados,* or slum dwellers, that made up roughly 10 percent of the city's population—an equally harsh existence in the urban environment. Illness, exacerbated by poverty, had taken its toll on Flavio, the eldest da Silva son. At night or when the smoke of the open cooking fire filled the da Silvas' six-by-ten-foot home, Flavio would succumb to violent coughing. With heaving chest, blue-tinged skin, and throbbing veins, his whole body would be consumed in the fight to breathe. This fight with bronchial asthma had left a visible mark on the boy; when doctors at CARIH in Denver saw Flavio in *Life,* they recognized his expanded chest and knew its cause (Figure 29). Convinced that in their institution the boy could be saved from the death that would soon meet him in his Rio slum, CARIH doctors wrote to the magazine's editors and offered free treatment. Other concerned and inspired Americans also reached out, sending hundreds of letters and donations to *Life* to help rescue Flavio from a life without hope.[1]

Gordon Parks, *Life*'s first African American photographer, had been sent to Brazil in March 1961 to capture in pictures and words the tragedy of poverty in Latin America. His article, "Freedom's Fearful Foe: Poverty," focused on the life of Jose da Silva, his wife, and their eight children. It was the second of a five-part series entitled "Crisis in Latin America," which *Life*'s editorial board had planned for that spring. Fearful that poverty, widespread illiteracy, and social injustice offered a fertile ground for Communist revolution, early in his administration President John F. Kennedy announced a $500 million plan—the Alliance for Progress—to end hunger, disease, and illiteracy in Latin America. While Kennedy's advisers struggled with how to implement their Cold War plans for universal education and economic development in the Western Hemisphere, Americans took comfort that summer in knowing that the compassion of a nation had rescued Flavio, a "symbol of impoverished millions," from the grip of want, disease, and despair (Figure 30).[2]

But while the nation seemed fixated on this Brazilian child, the weak cries of American children, suffocating under the weight of economic, environmental, and racial injustice, remained largely unheard. When a *Life en Español* version of the Parks article appeared, the Brazilian magazine *O Cruziero* was outraged by the hypocrisy in America's neglect of its own urban poor, and it sent reporters to document the harsh conditions of New York City's slums. Unfamiliar with the city's geography, the *O Cruziero* reporters searched the Wall Street district and chose a Puerto Rican family as their subject. Several staged shots later—a sleeping child covered in cockroaches, another crying in anguish from hunger—their work was done.[3] The reporters need not have gone to such lengths if they had only known where to look. In New York City, Chicago, New Orleans, and other American cities, conditions of despair approximating those of Catacumba were commonplace, even if a veil of inattention kept them hidden. As the Spanish Harlem writer and activist Piri Thomas observed, in America "There ain't no bright sunlight to reveal the stark naked truth of garbage-lepered streets."[4]

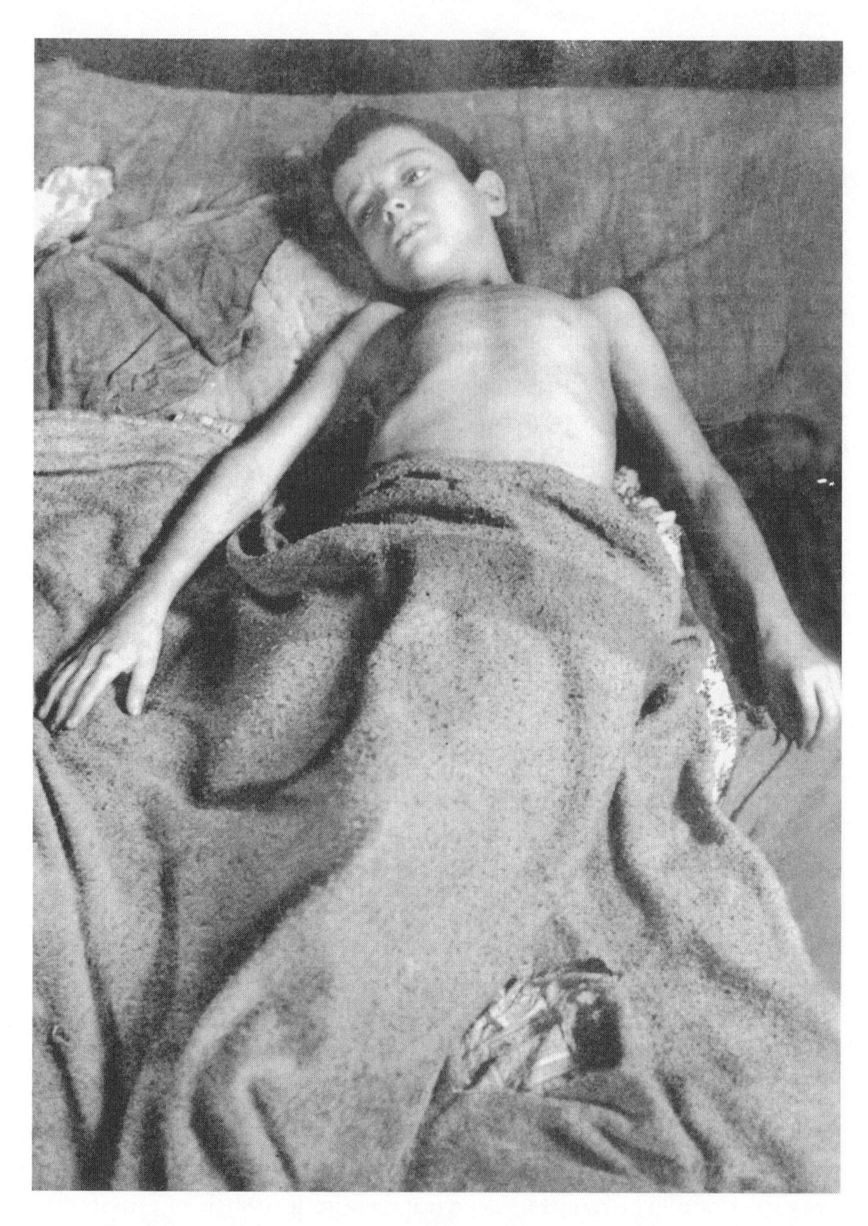

Figure 29. Bronchial asthma had left a visible mark on Flavio da Silva. Copyright The Gordon Parks Foundation.

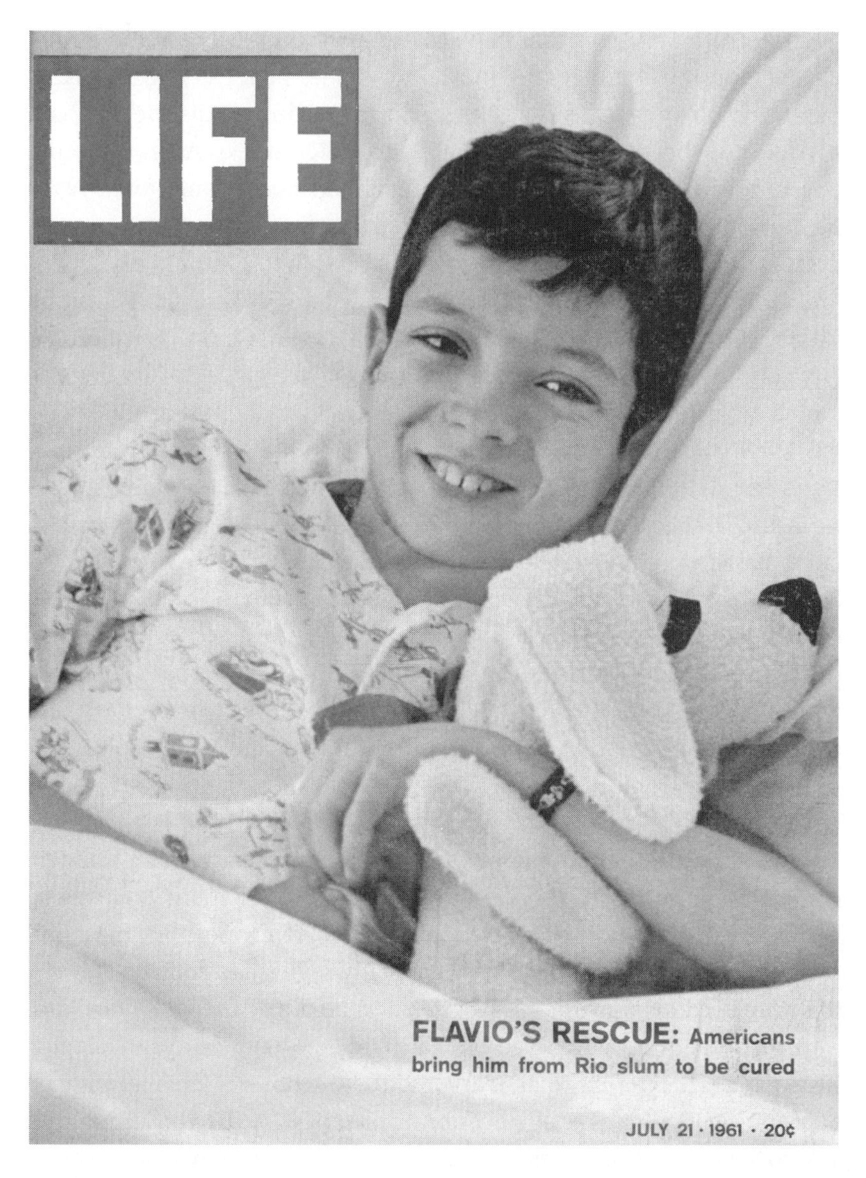

Figure 30. Flavio da Silva two weeks after his "rescue" from a Rio de Janeiro slum. Copyright The Gordon Parks Foundation.

The truth of America's urban ghettos—overcrowded conditions, decaying housing, high infant mortality, crime, poverty, and disease—began to well up in the early 1960s. So too did asthma, a disease that had mysteriously begun disproportionately to afflict African Americans and Puerto Ricans living in poor urban neighborhoods. Some forty years later, public health experts, physicians, and community activists expressed dismay at the alarming increase of childhood asthma morbidity and mortality as the twentieth century ended and the new millennium began, particularly among people of color living in impoverished inner-city communities. Among blacks and Latinos in New York City, for example, hospitalization and death rates are 3–5 times those of whites. In East (or Spanish) Harlem, where hospitalization rates for asthma are ten times the national average, an estimated 23 percent of children suffer from the disease.[5] To date, a host of factors—including genetic differences based on ethnicity, socioeconomic status, exposure to dust-mite and cockroach allergens, air pollution, and inadequate access to health care—have been cited to account for asthma's rise. But the roots of this urban public health crisis extend deep into our past, at least as far back as when Flavio first came to the United States to be cured. America's urban asthma epidemic, long in the making, is a product of the ecology of injustice that structures urban life.

Flavio's rescue in the summer of 1961 offered Americans a false sense of compassion and pride. White privilege had afforded many Americans the luxury of averting their eyes to the signs of poverty and sickness that existed within an economically and racially segregated society. Flavio's flight to freedom was not an experience shared by America's poor, minority populations, who lived increasingly in what black psychologist Kenneth Clark referred to as "dark ghettoes"—that is, "social, political, educational, and—above all—economic colonies" of the United States.[6] The social unrest bred by poverty that President Kennedy feared would push Latin Americans into Communist revolution was gathering momentum in the United States, taking form as a fight against economic and racial injustice. In May 1961, soon after a Supreme Court ruling barring segregation on interstate transit, the Congress of Racial Equality decided to test Kennedy's campaign commitment to civil rights. In Washington, D.C., whites and blacks boarded Greyhound and Trailways buses headed to New Orleans. Black "Freedom Riders" would sit in the front of the bus or wait in white-only seating areas in bus terminals. White activists would occupy public spaces reserved for blacks. Unlike Flavio's journey, the Freedom Riders' trip was met with violence, not compassion and love. In Aniston, Alabama, a white mob of more than two hundred people slashed the tires of one bus and then firebombed it. In Birmingham, an angry white mob met another bus and many Freedom Riders, both black and white, were severely beaten. The Freedom Riders were undeterred by the violence that greeted them. Armed with fresh recruits from Nashville, they pushed through Alabama, but only with the intervention of Attorney General Robert Kennedy and civil rights leader Martin Luther King, Jr., as well as the protection of state police and the Alabama National Guard. Mississippi proved the final roadblock, however, as over three hundred Freedom Riders were arrested by local police and tried by local courts. They never made it to New Orleans.[7]

In New Orleans, a less visible but no less disturbing symptom of racial and economic inequality made its seasonal appearance that fall. On 19 October 1961, the *Times-Picayune* reported an outbreak of asthma in the city. For two consecutive days, the emergency clinic of Charity Hospital was inundated with more than one hundred people per day seeking treatment.[8] Over 90 percent of the patients were African American, and the majority lived in communities that radiated outward from Claiborne Avenue, the thriving main street of New Orleans's black community and the main thoroughfare of the city's largest African American Mardi Gras parade. The epicenter of the epidemic was Treme, a historic black neighborhood founded by "free people of color" in the nineteenth century, just outside the French Quarter and adjacent to Charity Hospital (Figure 31). Run by the state of Louisiana, Charity Hospital was the largest branch of the state's free medical care system and limited treatment to the city's economically disadvantaged. Care extended to all the urban poor, but in the 1960s Jim Crow laws still ensured that indigent patients were segregated into separate "colored" and "white" wards. Although New Orleans was one of the least racially segregated American cities, the backswamp area, where the asthma epidemics were endemic, had the greatest concentration of blacks and poverty. In New Orleans, the average median income of blacks was less than half that of whites. At

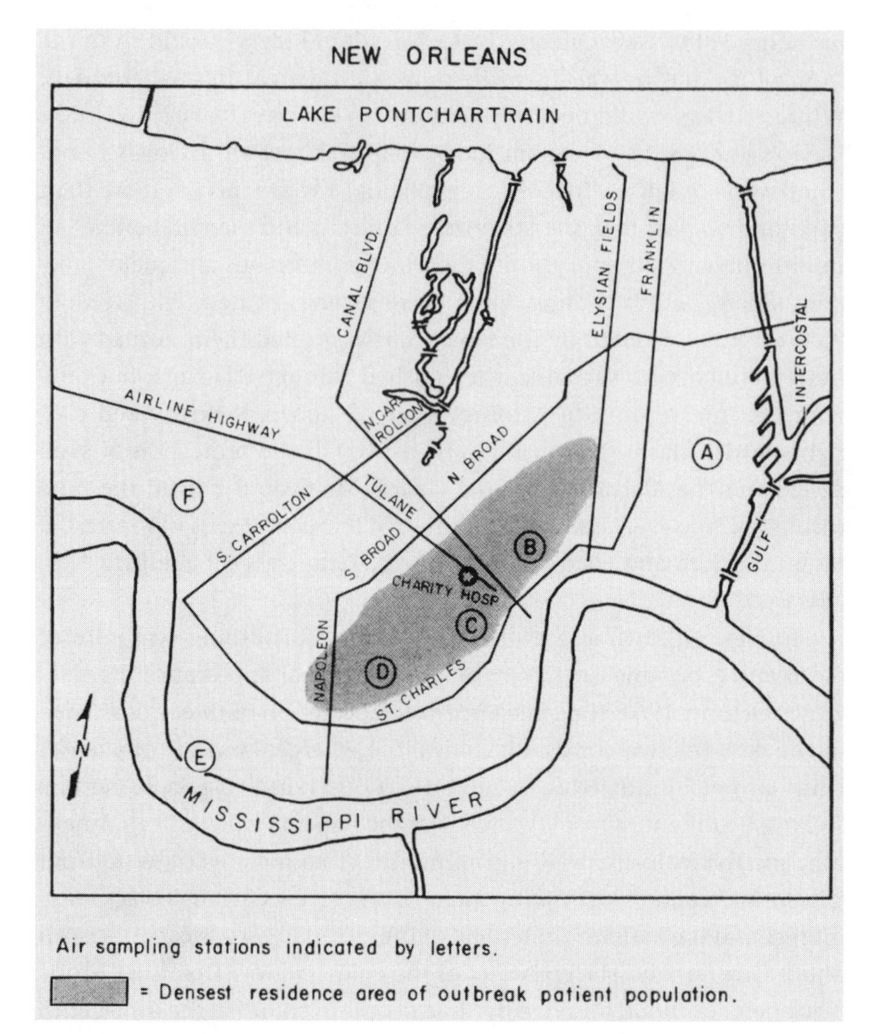

Figure 31. A map of the 1961 New Orleans asthma epidemic, showing Charity Hospital and the historic black neighborhood of Treme at the epicenter. Hans Weill et al., "Epidemic Asthma in New Orleans," *Journal of the American Medical Association* 190 (1964): 812.

Charity Hospital, 75 percent of patients admitted annually were African American, a consequence of historical patterns of settlement, as well as economic and racial segregation.[9]

The outbreak of asthma in the fall of 1961 among poor, African American communities in New Orleans was nothing new. It appeared so regularly that local New Orleanians referred to the annual event as the fifth season.[10] By 1961, it was barely a news story. And after Charity Hospital changed its policies and began charging patients for asthma drugs because of rising demand, advertisements appeared every fall in the city's newspapers for drugs like AsthmaNefrin, an over-the-counter bronchodilator.

Three years earlier, in August 1958, officials from the U.S. Public Health Service and the affiliated Robert H. Taft Sanitary Engineering Laboratory in Cincinnati, Ohio, had descended on New Orleans. They came after local reports of an asthma epidemic that resulted in three deaths had made national news. Beginning in the mid-1950s, an unusual annual pattern of spikes in asthma admissions to Charity Hospital appeared each summer and fall. October and November were the worst months. On some days, more than two hundred patients, many of them under the age of thirty, streamed into the emergency room for the first time, struggling to breathe. On normal days, the average number of asthma admissions was twenty-five.[11]

Public health officials working for the federal government had not seen such an epidemic of respiratory distress in the United States since 1948. That year, on a Tuesday morning in late October a thick fog had hung over Donora, Pennsylvania, a steel mill town located in a valley on the banks of the Monongahela River, southeast of Pittsburgh. By afternoon, the fog had turned into a "motionless clot of smoke," and the "air began to have a sickening smell." Meanwhile, the Donoran Zinc Works, which employed fifteen hundred of the town's fourteen thousand residents, continued operations, spewing thirty-six tons of sulfur compounds a day. Much of these emissions remained trapped in the valley. As the weekend approached, there was no sign of relief. Half the town became ill. As demand for oxygen surpassed the capabilities of the local hospital, volunteer firefighters made makeshift oxygen tents with oxygen canisters and blankets in people's homes. Those with histories of asthma or other respiratory diseases were particularly vulnerable. By the time the Donoran smog finally lifted the following Sunday, twenty people had died and thousands more had become ill. Although the U.S. Public Health Service's Division of Industrial Hygiene absolved the Donoran Zinc Works of any responsibility, the Donoran smog came to symbolize the ill effects of postwar economic and urban growth, and a new era in federal research was launched on the long-term health effects of air pollution.[12]

The cases of asthma in New Orleans, however, were unlike those in Donora or any other major episodes of acute breathing difficulty that descended in the 1950s upon those living in major metropolitan centers such as Los Angeles, New York, London, and Tokyo. In early December 1952 during London's Great Killer Fog, for example, hospitals overflowed with people whose airways were choked by a deadly smog that hung over the city. Mortality rates soared as the mist of black coal smoke thickened into pea soup, shrouding the city in twilight day after day. After several weeks the phantom epidemic and mysterious fog vanished, never to return. But the New Orleans asthma outbreaks proved both regular and predictable. Asthmatics had learned to recognize the signs of the epidemic based on the time of year, climatic conditions, prevailing winds, and geographic location. At first, public health officials searching for the cause of the New Orleans epidemic thought that a form of bronchial asthma familiar to physicians and military personnel living in the Tokyo-Yokohama area of Japan might offer clues. From September through May, a significant number of personnel stationed in this heavily industrialized region would develop a night cough that grew progressively worse. Severe wheezing and shortness of breath ensued.[13] Although the localization of Tokyo-Yokohama asthma to a small geographic area had parallels to the New Orleans outbreaks, the similarities ended there. While many of the Tokyo-Yokohama patients had no prior history of asthma or allergy, over 99 percent of those in New Orleans were either allergy sufferers or responded positively to a standard allergen skin test. The majority of military personnel affected in Japan were smokers. In New Orleans, much to the surprise of physicians, only a relatively small percentage of the patients smoked. Few had signs of emphysema or a history of chronic bronchitis. Furthermore, unlike Tokyo-Yokohama patients, all New Orleans asthma victims responded well to standard emergency room treatments for asthmatic attacks. To researchers, the New Orleans epidemics seemed a clear-cut case of allergenic asthma, uncomplicated by other respiratory diseases.

But the seasonal epidemic continued to baffle investigators from Tulane University's School of Medicine and School of Engineering, hired by the U.S. Public Health Service to track down its cause. Robert Lewis, a biostatistician who headed the Tulane team, assumed that a single or perhaps multiple-point source of allergenic air pollutants was responsible for these outbreaks. To isolate possible sources, air-monitoring stations were placed at six locations within a three-mile radius of Charity Hospital. A team of twenty public health nurses and an air-sampling crew were also assembled. A few hours after notice from the hospital staff, the team went into action. The nursing group undertook extensive interviews of asthma patients in Charity's Emergency Center, while the air-sampling team gathered time-sensitive data on wind direction and speed, temperature, humidity, particulate matter concentration, and other meteorological phenomena.

The mystery seemed to be solved in 1962, when Lewis's team found a significant correlation between seasonal asthma attacks and calls to the city fire department reporting fires in nearby dumps.[14] The New Orleans asthma zone was surrounded by three of the city's largest landfills. Simmie Harvey, state president of the Southern Christian Leadership Conference, the civil rights organization whose local branch had fought to clean up former incinerator sites in New Orleans, recalled that as a youth these "sites loomed over the poor of our city and the smell was often sickening."[15] Since 1909, the city had used a ninety-five-acre site located on Agriculture Street as a city dump. By the 1960s, the site had been the collection point and dumping ground for the city's garbage, commercial waste, and incinerator ash for more than forty years. And it sat less than a half mile from the residential area with the greatest density of asthma outbreaks. In 1948, nearby residents began to complain of offending odors. Four years later, the Agriculture Street dump was shut down and turned into a sanitary landfill.[16] In 1994, it was designated an EPA Superfund site. But underground fires became a common sight at the dump a decade after it was shut down. Refuse, waste products, and toxic chemicals spontaneously ignited when the water table receded during the drier months of the year. Fissures opened in the ground, releasing smoke plumes that carried both particulate matter and gaseous fumes containing countless unidentified substances. Local residents nicknamed the dump "Dante's Inferno." Subterranean fires also burned a few miles northeast of the Agriculture Street landfill at the Gentilly Street dump. To the southwest, unidentified, highly allergenic and toxic particulate matter emanated from a dump built on a batture (the alluvial land between the low water level of the Mississippi and a levee) near Audubon Park. When winds blew up from the Gulf of Mexico, these particulates swept across New Orleans's backswamp ghetto. Although the precise allergenic substance responsible for the asthma epidemics could not be identified, the Tulane research team determined, based on skin testing, that a poorly combustible, silica-containing particle isolated from the plumes of "Dante's Inferno"—the Agriculture Street landfill—was a likely suspect.[17] In response, the city removed three hundred thousand yards of excess fill from the Agriculture Street dump in an effort to squelch the smoldering fires and episodic epidemics.

But the fall asthma epidemics did not subside as the garbage fires burned out. Determined that a single allergenic air pollutant or possibly a combination of pollutants was responsible for New Orleans's fifth season, investigators from Tulane and the U.S. Public Health Service zeroed in on a large public grain elevator along the Mississippi River, southwest of Charity Hospital, as the next likely suspect. As the nation's largest grain port in 1964, New Orleans exported over 254 million bushels of wheat, corn, and soybeans. As grain was moved from elevator to ship, grain dust filled the air. Low-velocity southwest winds readily carried the grain dust to neighboring communities, where asthmatic residents reacted to the allergen.[18] The grain elevator hypothesis, however, could not readily explain the precise seasonal nature of the epidemics. Other researchers looked to fluctuating atmospheric concentrations of mold and pollen counts to account for the ongoing public health crisis.[19] At the time, however, no one paid any attention to how the ecology of injustice in the city affected New Orleans's underclass population.

Epidemiologists, physicians, and engineers hoped to explain New Orleans asthma on the basis of a single chemical pollutant, industrial airborne particle, or natural allergen. For a single cause there would be a single technological solution. Americans had come to expect technological quick fixes of medicine in the 1950s. After all, "wonder drugs," so it seemed, had already eliminated deadly and crippling infectious diseases such as polio, smallpox, and tuberculosis. But the quest for a magic bullet to rid the city of epidemic asthma was a quixotic search. It was a diversion that obscured the harsh inequities that left people vulnerable to a disease that elsewhere was less pervasive and less life-threatening. Economic, environmental, and racial inequities were all at play in New Orleans asthma. Particulate matter spewing from the city's major landfills settled in the lungs of the nearest residents, those who lived in the city's poorest and least racially mixed neighborhoods. These low-lying neighborhoods also happened to be near urban industrial districts.[20] Location and poverty, coupled with segregation, contributed to poor air quality and to poor housing quality as well. Dilapidated, unair-conditioned wooden houses were a common sight in the central city until the late 1960s, when inner-city urban renewal began. People living in such homes were exposed to more outdoor airborne allergens than those who could afford to insulate themselves from the city's dirty air. Health care inequities took a toll as well. Poor African Americans did not have access to the regular medical care available to middle-class whites, whose asthmatic attacks were most often less frequent and less deadly. Only in 1966, with the establishment of Medicaid, did indigent patients gain the possibility of regular care from a private physician in addition to free emergency room treatment.

In the early 1970s, John Salvaggio, a clinical physician at the Louisiana State University School of Medicine and Charity Hospital, suggested that perhaps there was nothing anomalous about New Orleans asthma. For almost a decade, investigators had sought to explain these mysterious epidemics on the basis of some unique environmental factor limited to New Orleans. But Salvaggio began to think otherwise. Perhaps, Salvaggio argued, Charity Hospital's experience with asthma admissions had merely opened a window onto the natural history of a disease; perhaps asthma targeted urban centers and its indigent populations, which went "from crisis to crisis without continuity of medical care."[21] Asthma had begun to emerge in the urban ecology of America's inner cities as a disease symptomatic of the disparate spaces in which Americans lived and breathed. New Orleans asthma was just the beginning of a growing national public health crisis.

In the fall of 1962, Dr. Leonard Greenburg noticed a peculiarity in the pattern of emergency clinic admissions for asthma that were reported by New York City hospitals. Greenburg had spent the past decade looking for convincing evidence of a link between air pollution and the higher rates of illness and death in New York City. As the first commissioner of New York City's Department of Air Pollution Control, he had been relatively ineffective in containing the city's growing air pollution menace. Even when a six-day siege of smoke, haze, and smog paralyzed the city in November 1953—shutting down LaGuardia Airport, crippling commuter traffic, and halting bus service—Greenburg was unable to mobilize city officials to take action.[22] Despite sulfur dioxide concentrations that far surpassed those of London's 1952 killer fog and regardless of increased hospital admissions, Greenburg could offer little proof at the time that air pollution jeopardized the health of New York City's residents. Without sufficient funds, staff, political will, or conclusive scientific data, Greenburg could do little but watch as major polluters (such as the electric company Consolidated Edison) and numerous apartment house incinerators added to the thousands of tons of particulate matter, sulfur dioxide, and other air pollutants each year. When he resigned as air pollution commissioner in 1960, Greenburg devoted his career to finding the smoking gun that he had lacked as chief watchdog and regulator of New York City's air quality.[23] Many physicians assumed that a relationship existed between air pollution and asthma, suspecting that air pollution contributed to increased sickness and death. To test this assumption, Greenburg combed the records of New York City hospitals during periods of high levels of air pollution, looking for a corresponding increase in asthma patient admissions. But Greenburg found no such correlation. What he did find was far more puzzling and surprising.[24]

Increasing numbers of asthma patients had been inundating emergency rooms in New York City over the previous decade with no obvious correlation to days of poor air quality. Among four New York City hospitals, Greenburg and his colleagues found a two and one-half to eightfold increase in the number of asthma visits to emergency clinics from 1952 to 1962. No one knew the cause. But two hospitals stood out: both were in upper Manhattan, and both served poor, minority communities. At Harlem Hospital, located on Lenox Avenue between 136th Street and 137th Street, one out of every four visits to the emergency room, excluding trauma and obstetric visits, was for asthma. At Metropolitan Hospital, located on East 97th Street and Second Avenue, one out of every seven visits was for asthma. Furthermore, a marked increase in asthma admissions occurred each fall. New York City was beginning to look a lot like New Orleans.[25]

The environmental conditions endured by the African American community served by Harlem Hospital were as bad as those faced by the asthma patients of New Orleans. The great migration of blacks from the rural South to northern industrial centers after World War I had turned Harlem into a vibrant center of black literary and artistic life during the 1920s, and the migration continued apace after World War II. Fleeing Jim Crow and a depressed agricultural economy, 1.5 million African Americans left during the 1950s what Kenneth Clark described as the "miasma of the South, where poverty and oppression kept the Negro in an inferior caste."[26] Seeking educational and economic opportunities in the North, blacks encountered instead the harsh realities of an urban environment that was isolated spatially, socially, and economically from the goods, services, and employment that sustained the health of the city and the majority of its white residents.

Of the 240,000 residents who lived in 1960 in Central Harlem (a three-and-one-half square mile area lying between 110th Street to the south and the Harlem River to the northeast) 94 percent were African American. Limited largely to low-paying, unskilled, or semiskilled service jobs, Harlem residents earned one-third less than the average New Yorker. A severe housing shortage, coupled with inflated rents, resulted in overcrowded and dilapidated housing conditions that were unparalleled elsewhere in the city. Decaying tenement buildings, built before 1929, provided the bulk of available housing. Slum landlords showed up when high rents were due but were absent when renters pressed for repairs. More than half of the housing units in Harlem were classified in the 1960 census as unsafe, inadequate, or in need of major repair. Many units lacked heat or plumbing; often apartment dwellers relied upon gas ovens as their only source of warmth in cold winter months. "Rats and roaches," one resident living on West 117th Street complained, "were literally moving the tenants out of the building." More often than not, the Department of Health failed to prosecute landlords for housing code violations.[27]

Such conditions, exacerbated by the discriminatory policies of the city's social and health agencies, led to significant health disparities. The infant mortality rate in Central Harlem was nearly double that of New York City as a whole in 1961. Tuberculosis afflicted twice as many Harlem residents as it did those living in the rest of the city. Poverty, poor housing, and inadequate access to medical care were just a few of the factors that contributed to a population vulnerable to diseases that were less threatening under other economic and ecological circumstances. In the 1960s, few middle-class suburban whites were likely to view asthma as a deadly illness. But nearly half of New York City's poor black welfare mothers surveyed in 1966 considered asthma to be a very serious disease.[28]

Similar conditions of poverty, degraded housing, and poor quality health care prevailed in the neighborhood served by Metropolitan

Hospital, where Greenburg's team found the second highest rate of hospital asthma admissions in New York City. Located on the southern edge of Spanish (East) Harlem, the heart of New York City's Puerto Rican community, Metropolitan became the primary health care provider for the city's poor Hispanic population. In the prosperous postwar economy of the United States, many Puerto Ricans saw migration as a path to upward social mobility. Of the roughly eight hundred thousand Puerto Ricans who came north for low-paying jobs in the blue-collar trades and manufacturing industries between 1940 and 1960, almost 80 percent flew from San Juan to New York City.[29] Many expected to return to their beloved *isla verde* once their economic future was secure. But "los Estados Unidos," observed Dolores Montanez, could be a "cold place to live—not because of the winter and the landlord not giving heat but because of the snow in the hearts of people."[30] In Spanish Harlem, the median income was roughly equivalent to that of Central Harlem and far below that of other New York neighborhoods. "Old-law" tenement buildings, dating back to the nineteenth century, were common in East Harlem. In these five- or six-story buildings, families lived in crowded quarters; three or more people often occupied a single room that lacked adequate ventilation and sunlight. Most tenement apartments of this vintage had no separate baths. The toilet was housed in a closet. The tub was located in the kitchen-dining room area and often doubled as a utility table. Some apartments lacked central heat. Rent strikes were the only recourse tenants had in drawing the city's attention to the inhumane conditions. Landlords did little to respond to the violation notices city inspectors issued. Rates of respiratory infections and tuberculosis were high in these crowded conditions. A survey undertaken in the 1950s of eighty Puerto Rican families in East Harlem showed that 14 percent had a member of the household with chronic bronchial asthma. In the majority of cases, the family member was a child whose symptoms had developed after immigration.[31]

In the early 1960s, the hot zones of urban asthma in New York City were geographic areas with high concentrations of African American or Puerto Rican families, many of whom had come north after World War II. Both epicenters of these concentrations—Central Harlem and Spanish Harlem—also had the greatest concentration of poverty in New York City and some of the highest population densities per square mile in the world. Epidemiologists puzzled over whether population or environment could account for the unprecedented rise in hospital admissions for asthma. Events that transpired in Harlem and throughout the nation, however, quickly transfixed attention upon race as the determining factor.

On a hot summer evening in July, just two weeks after President Lyndon B. Johnson signed the Civil Rights Act of 1964 into law, thousands of blacks in Harlem, many of them teenagers, took to the streets in anger and protest. It was neither the beginning nor the end of a series of long, hot summers that saw riots and devastation in Chicago, Cleveland, Detroit, Newark, and Los Angeles. Frustration and anger that set ghettos burning would later be channeled into calls for Black Power; out of the ferment of youth came the energy that tore down and built a new image of the black ghetto, one founded upon self-help, community action, and racial pride.[32]

While the black community transformed anger and rage into a positive force of social change, white psychiatrists and clinical allergists looked upon such emotions as the pathological seat of the nation's rising asthma epidemic. In late July 1965, irate Watts residents and a prickly Los Angeles police force and National Guard clashed in a firestorm that left thirty-four people dead and hundreds injured on the West Coast. The events prompted *New York Times* reporter John Osmundsen to popularize the first of several explanations for the sharp rise in asthma among New York City's African American and Puerto Rican populations. "An emotional epidemic that has probably never been paralleled in the annals of medicine" was sweeping the urban ghetto, Osmundsen wrote. Medical authorities suspected, he reported, that "tensions arising from the civil rights movement" were its likely cause.[33]

The impulse to invoke race to explain the increase in urban asthma first witnessed in the early 1960s speaks both to the popularity of psychosomatic explanations of asthma during the period and to the medical profession's long and ongoing history of looking to racial difference to account for observed health disparities. Osmundsen acknowledged that other factors, including air pollution, housing, socioeconomic status, and geography, might be involved in inner-city asthma. But he noted that whatever role such factors played, they were "doing so against a background of emotional conflicts which have been reported widely in medical literature to be associated with asthma attacks."[34]

Many psychiatrists regarded asthma as based upon deep-seated emotional insecurities, coupled with an intense need for dependence. Nevertheless, the mechanisms by which the emotional process interacted with the respiratory, immunological, and central nervous systems remained obscure. According to psychosomatic theories of asthma, anger, which a child might repress for fear of losing his mother's affection, could provoke an asthma attack. Another provocation might be the conflict between a wish for independence and a felt dependency. In the case of white children, psychiatrists ascribed such emotional tensions to such factors as an overprotective mother, an explanation that factored into Murray Peshkin's parentectomy treatment at CARIH.[35] As asthma rates among African Americans and violence in American ghettos soared, some psychiatrists and clinical allergists working in the field of asthma seized upon the "black personality" as a likely cause of the disease in African American children and the increasing preponderance of asthma within the black ghetto.

The associations of asthma with the psychology of race owed much to postwar social science research, which painted a portrait of a black psyche damaged by centuries of racial prejudice and discrimination. In their highly influential 1951 book, *The Mark of Oppression,* Columbia University psychiatrists Abram Kardiner and Lionel Ovesey argued that self-hatred was a common trait of the black personality. Black children, seeing their parents as members of a "despised and discriminated-against group," aspired to be white. "Accepting the white ideal," these physicians argued, was a "recipe for perpetual self-hatred, frustration, and for tying one's life to unattainable goals." Hatred of white society was another mark of oppression that the psychiatric community saw manifested in the black psyche as uncontrolled rage. To liberal social scientists, the violence and aggression in the urban race riots of the 1960s were unhealthful. Not only did it lead, as Helen V. McLean at the Institute for Psychoanalysis in Chicago argued, to a greater number of health problems in the black community, but it also threatened liberal hopes for an integrated society by propelling the civil rights movement toward black nationalism.[36]

To epidemiologists and physicians, the epidemic wave of asthma among poor African Americans was symptomatic of the damaged black psyche and the pathology of racial prejudice in America. Osmundsen suggested that it was not hard to see how asthma, "precipitated by conflicts between hostile feelings and dependent needs, [could] arise among members of racial minority groups on whom civil rights activities focus."[37] Factors that psychiatrists used to evaluate the black personality also became a part of the clinical diagnosis of the black asthmatic child. Records of medical evaluation conferences at CARIH in Denver, for example, reveal that the psychology of race figured prominently in the evaluation of black children who were admitted. When Mary, an eight-and-a-half-year-old African American child from a poor neighborhood in Kansas City entered CARIH in 1959, part of the discussion for treatment centered on her alleged "feelings of racial inadequacy." "She believes that her color is bad and is obsessed with being different and inferior in color," wrote one attending physician. "She's struggling with this inside herself, but she hasn't really resolved this at all." Two years later, physicians regarded Mary's resolution of these racial conflicts, achieved through the help of play therapy, as one indication of her successful rehabilitation.[38]

To presume that asthma emerged from the psychological blight of a segregated society ignored the environmental inequities of the urban ghetto that were visible to African American and Latino residents. The ecology of injustice was evident everywhere in the deteriorating conditions of the inner city, where poor people of color lived. The distribution of allergens and pollutants was not equal in the economically and racially segregated spaces of the city. Neither was medical care. But the inability of epidemiologists and physicians to perceive such environmental inequities only points to how white privilege blinded medical authorities and prevented them from recognizing that the urban ecology of asthma was different from other breathing spaces where predominately white middle- and upper-class Americans lived.[39]

In the early 1960s, the ghetto was among the places least familiar to the clinical field of allergy. Allergists neither knew nor worked with nature in the urban ghetto. Consequently, they had few clues in their doctors' bags of skin tests and allergenic extracts to help identify important environmental factors that might be responsible for the increase of asthma observed within the city's poor African American and Latino communities. Pollen was the most obvious and best-known allergen. Smog, an integral part of urban city life, was also a likely suspect, although the scientific jury was still out on the role of air pollution in the development of asthma. Neither pollen nor air pollution, however, in either isolation or combination offered a compelling explanation. But a clue did emerge that helped shed light on this medical mystery.

Physicians had long been aware that the sting of a bee or other member of the Hymenoptera family could send an allergic patient into anaphylactic shock. Yet the role of insects in the onset of allergic reactions received far less attention than pollen, molds, and house dust prior to World War II. The rapidly expanding pest control industry in the postwar era, however, threw a searchlight on arthropods as potential allergens. Among pest control workers, entomologists, and people working in the silk and grain industries, where exposure to insects and their by-products was common, workers seemed particularly susceptible to certain kinds of allergic reactions. Merely the odor of the Madeira cockroach, for example, had caused an asthma attack in one laboratory assistant working in an entomological research lab. Two physicians in Washington, D.C., believed such evidence suggested that insects should be added to the list of known allergic triggers. To test their hypothesis, Dr. Harry S. Bernton and Dr. Halla Brown screened patients from their allergy clinics at Freedmen's Hospital, Providence Hospital, and George Washington University Hospital, as well as poor but healthy individuals in the surrounding population, for sensitivity to extracts derived from the American and Oriental cockroach. To their surprise, 28 percent of their allergic patients displayed a positive skin reaction to the cockroach blender mix. Among the healthy sample population, 7.5 percent showed a positive response.[40]

Bernton and Brown made no mention of ethnicity in their initial study and only passing reference to the socioeconomic background of their patients. But such details are not hard to infer. Freedmen's Hospital, established in 1862 for the treatment of ex-slaves moving to Washington, was the only federally funded hospital in the nation for the care and treatment of black patients. Providence Hospital, established by Abraham Lincoln in 1861, had grown into a nonprofit community hospital run by the Catholic Diocese, and it largely served the uninsured poor, which included an expanding immigrant Latino population. The racial and class backgrounds of the patients that Bernton and Brown screened for cockroach sensitivity appeared strikingly similar to those of individuals most affected by New York City's urban asthma. Could the increase in urban asthma among New York City's African American and

Latino populations be related to contact with cockroaches? With a grant from the John A. Hartford Foundation, Bernton and Brown attempted to find out.

Working with allergy clinics in seven hospitals in New York City, Bernton and Brown tested 589 patients for sensitivity to an extract derived from the body parts of the German cockroach, *Blattella germanica,* the most common urban pest. This time, however, the two physicians classified patients and their responses according to four ethnic groupings—Puerto Rican, Negro, Italian, and Jewish. Bernton and Brown based the design of their study on an unpublished survey of cockroach infestations in New York City slums; it allegedly found the dwellings of Puerto Rican and "Negro" families to have the severest infestation rates, while the homes of Italian and Jewish residents had fewer cockroaches. In the entomological survey, ethnicity rather than class served as the primary category of analysis; differences in median income or housing were left unexamined. Among the patients screened in public clinics, Bernton and Brown found that 59 percent of Puerto Ricans and 47 percent of African Americans were sensitive to cockroach allergen. In contrast, only 17 percent of Italians and 5 percent of Jews reacted positively. The order of cockroach sensitivity by ethnic group corresponded almost identically to the severity of cockroach infestations found by the cockroach survey. Furthermore, 63 percent of the patients involved in the study were reported to be asthmatic. These findings, Bernton and Brown argued, emphasized the need for urban allergy clinics to make screening for cockroach sensitivity a standard part of clinical diagnosis and to administer corresponding desensitization treatments for allergy and asthma sufferers.[41]

Although Bernton and Brown acknowledged class to be a confounding factor in the differential exposure of ethnic groups to cockroaches, such subtleties were easily lost in the popular press. Furthermore, in drawing attention to the heritability of asthma, their work made it easy to infer that differential asthma rates might be explained according to biological racial differences. Just weeks after Bernton and Brown published their findings in the *Southern Medical Journal,* the *New York Times* announced in early September 1967 that allergy to cockroaches had been identified as a possible cause of the "startling rise in recent years in the incidence of asthma among New York Negroes and Puerto Ricans."[42]

Cultural attitudes toward the cockroach played heavily into epidemiological assumptions that highlighted race rather than class as the primary determinant in understanding the New York City outbreaks of asthma. Racial assumptions about the sanitary habits of ethnic minorities figured prominently in explanations of why cockroach infestations—and therefore exposure—were most prevalent in the inner city. "By all odds," Robert Stock wrote in a *New York Times Magazine* article based in part on Bernton and Brown's findings, the cockroach was "the most reviled creature on the face of the earth."[43] Dirt and danger were the marks of its reputation. Cockroaches, the U.S. Department of Agriculture warned, "carry filth on their legs and bodies and may spread disease by polluting food."[44] Pest control manuals targeted them as the most "repulsive" of urban pests.[45] To call someone a cockroach was to invoke the vileness of this most despised insect. It was a common racial slur leveled against New York City's immigrant Puerto Rican population. After the Jets first clash with the Sharks in the 1961 film version of *West Side Story,* the white teen gang members cast insults upon the Puerto Ricans moving into the neighborhood: "They multiply. They keep comin'. Like cockroaches. Close the windows. Shut the doors. They're eatin' our food. They're breathin' all the air."[46]

But to downtrodden African Americans and Latinos, cockroaches had nothing to do with their cleanliness or behavior. Rather, cockroaches spoke to the despicable ways of slumlords, who turned the ghetto into a colony of white America. To the residents of Central Harlem and Spanish Harlem, the cockroach was linked, not to race, but to the inhumanity of beings toward one another. In their eyes, the cockroach became an ally in the protest against economic, racial, and social injustice.

Contrary to popular opinion, cockroaches possess few filthy habits. They are in fact the felines of the insect world, constantly grooming their antennae, which they use to detect precious water and food resources (Figure 32). In the tropics, where the majority of the estimated four thousand cockroach species abound, the relatively constant warm temperatures and high humidity are paradise to this water-loving, heat-craving insect. But cockroaches have been on this planet a long time—Thomas Henry Huxley, the famous nineteenth-century British biologist, believed them to be the archetypal insect—and throughout their evolutionary history they have managed to adapt to virtually every ecosystem

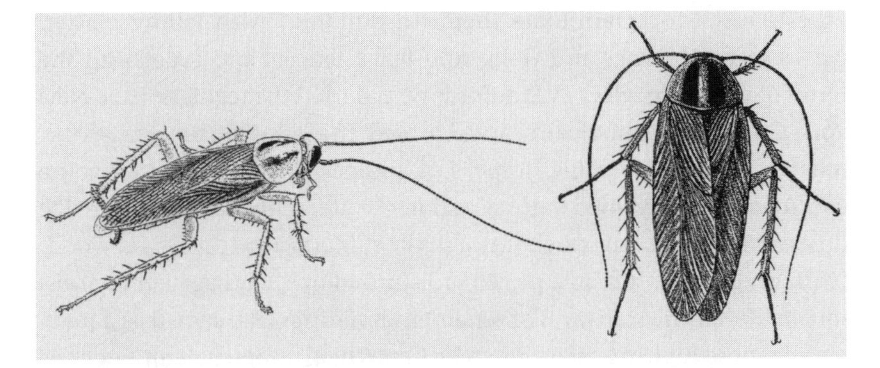

Figure 32. The German cockroach (*Blattella germanica*). Courtesy of Dr. William Robinson.

on Earth. Less than 1 percent of the known species have come into close contact with humans. Of these, the German, American, and Oriental cockroach are the most common domestic species. None of them are endemic to North America; they came from tropical Africa and East Asia, successfully colonizing the temperate urban landscape.[47] In torrid climates, exposure to these allergenic species is directly related to social status. In the Dominican Republic, for example, scientists have found that among poor children inhabiting drafty, wood-frame homes, where moisture sources such as toilets and sinks are generally located outside the living quarters, the incidence of cockroach sensitivity is low. In these shanties, the ideal conditions for cockroaches—high humidity and limited air movement—are lacking. In contrast, the tight masonry construction characteristic of better-built homes eliminates rapid air exchange and creates pockets of dead air with higher humidity—virtually a five-star hotel for the ubiquitous urban cockroach. The incidence of cockroach allergy among Dominicans in better-quality homes is significantly greater than among those who live in the shanties.[48]

In northern American cities, however, a different set of environmental factors—human and natural—conspired to create a quite different ecology of human-cockroach interactions. Old tenement apartment buildings, which made up the majority of housing stock in Central Harlem in the 1960s, were fertile breeding grounds for *Blattella germanica* and its domestic cousins. Cockroaches were a nightmare, not only for the residents of urban, low-income apartments, but also for pest control experts hired to exterminate them. In buildings with falling plaster, cracks in the ceilings and walls, and leaky faucets and toilets—all the result of years of neglect by landlords who skirted the regulations of New York City's Board of Health and ignored the pleas of tenants—cockroaches found a welcome home. Leaky pipes offered a source for the cockroach's most valued natural resource: water. Their antennae are like dowsing rods that can search through the subterranean caverns of wood, brick, and mortar to find a prized moisture source. Garbage left to accumulate under the stairways of tenement buildings also served as a magnet for these highly gregarious creatures, which congregate in kitchens and other food outlets, coming together, like humans, for conversation and a meal. The German cockroach is also highly mobile. Plumbing connections, heating and venting systems, and electrical conduits all serve as express highways for the frequent migration of cockroaches from one apartment to another to escape pesticides or to find better food and water supplies. The much higher cockroach populations in lower socioeconomic communities were a function not of race but of deteriorating housing conditions, which provided a perfect ecological niche for this preeminent survivor of the insect world.[49]

When Gordon Parks turned his camera on the American ghetto in 1968, he found conditions of poverty that equaled those of any Rio de Janeiro slum. In his *Life* essay on the Fontenelles, a black family struggling to survive in Harlem, Parks confronted white middle-class readers with a different America but of their own making. The Fontenelles lived in a "building ain't fit for dogs" and owned a cat "to keep the roaches and rats in check" (Figure 33).[50] While the Brazilian magazine *O Cruzeiro* had pasted cockroaches on the face of a Puerto Rican child in its fabricated story of New York City poverty, cockroaches crawling on the faces of sleeping infants were a common part of life in New York City slums. The *New York Amsterdam News,* the black newspaper of Central Harlem, regularly featured advertisements for rat and roach pesticides like "Kill Jo Paste" next to Primatene tablets, an over-the-counter asthma medication (Figure 34). In *The Cool World,* Warren Miller's gritty 1961 novel of life in Harlem, only "E Z Kill roach powder" and "Kill Kwik rat pellets" lined the shelves of the local hardware store.[51] By the 1960s, epidemiological evidence and everyday experience had confirmed that asthma and roaches had become part of the urban ecology of America's choking cities. But to isolate cockroaches as the cause of the first epidemic wave of

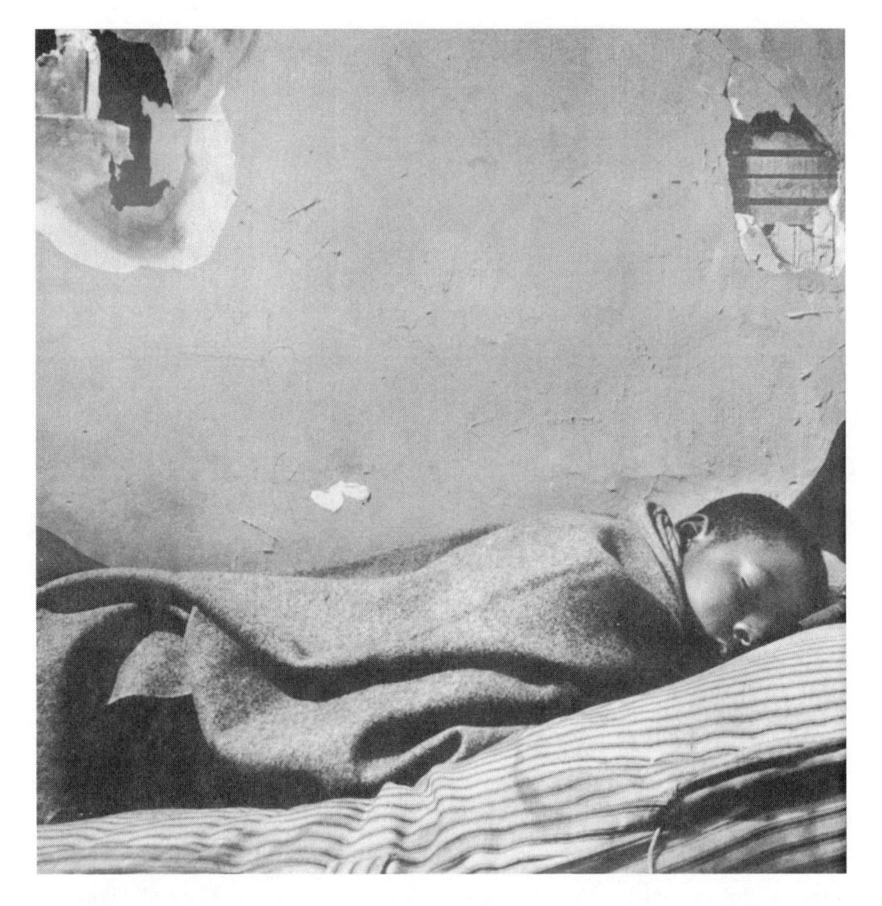

Figure 33. Gordon Parks's photographs of the Fontenelle family, such as this one, confronted *Life* readers with the hunger, poverty, and despair of America's urban slums. Copyright The Gordon Parks Foundation.

urban asthma would be to miss the larger environmental inequities at work in the ghetto; these inequities combined to make conditions ripe for the spread of an emerging disease and an opportunistic insect.

Race riots, not asthma, were what finally brought national attention to the poverty, poor housing, and lack of quality medical care that prevailed in the ghetto and contributed to the urban ecology of a disease. The Economic Opportunity Act, passed by Congress in August 1964, just weeks after the country looked with trepidation upon the summer firestorm in Harlem, formed the backbone of President Johnson's "War on

Asthma and Hay Fever Relief Comes in Minutes ...and Lasts For Hours

Tiny Tablet Now Available Without Prescription!

New York, N. Y. (Special) — Medical Science has developed a new, tiny tablet that not only stops asthma spasms, but brings relief to those who suffer from hay fever attacks.

Authoritative tests proved this remarkable compound brings relief in minutes — and gives hours of freedom from recurrence of painful spasms.

This fast-acting formula is prescribed by doctors for their private patients who suffer from asthma or hay fever. And now sufferers can obtain this formula — *without prescription*—in tiny, easy-to-take tablets called *Primatene*.®

Primatene opens bronchial tubes, loosens mucous congestion, relieves taut nervous tension, helps dry up nasal passages. All this without taking painful injections and without the inconvenience of nebulizers.

The secret is — Primatene combines 3 medicines (in full prescription strength) found most effective in combination for asthma and hay fever distress.

So look forward to sleep at night and freedom from asthma or hay fever spasms ... get Primatene, at any drugstore. Only 98¢ — money-back guarantee.

© 1957 Whitehall Pharmacal Company

Figure 34. Advertisements for Primatene tablets and Kill Jo Paste from the *New York Amsterdam News* illustrate the extent to which asthma had become endemic in Harlem by the late 1950s. *New York Amsterdam News,* 19 September 1959, 6.

Poverty." An infusion of funds over the next eight years into existing and newly established federal programs—including Aid to Families with Dependent Children, food stamps, Head Start, Medicare, and Medicaid—helped to lessen the toll of poverty—hunger, poor health, economic insecurity, and lack of opportunity—that faced an estimated 10 million Americans living in urban areas in 1964. When the Kennedy administration had first turned its attention to poverty in America, rural Appalachia, not the inner city, had been the place foremost in its mind. But the incendiary racial violence that erupted and spread across America's urban landscape in the summers of the mid-1960s fixed the eyes of the White House on another America. The need to secure the black vote for the Democratic Party in the upcoming 1964 reelection campaign likely played a part as well. In looking to the root cause of racial disorder, Washington officials found themselves confronting deep structural economic and social inequalities. When the Economic Opportunity Act passed, just two months after the Civil Rights Act became law, the federal government put poverty and race at the center of efforts to address the problems that plagued America's cities and stirred social unrest.[52]

New York City proved an important testing ground for the future direction of federal antipoverty programs. In the early 1960s, Richard Cloward, a professor at the Columbia School of Social Work, helped to create Mobilization for Youth (MFY) on Manhattan's Lower East Side. A pilot program in combating poverty, MFY was built upon principles of opportunity and empowerment. Lack of access to education and employment and the absence of political power had locked inner-city youth into a cycle of poverty and despair. This was the argument Cloward and his Columbia colleague Lloyd Ohlin advanced in their highly influential book, *Delinquency and Opportunity,* which captured the interest of Kennedy's circle of advisers. With funds from the President's Committee on Juvenile Delinquency, Cloward and others put theory into practice and built MFY into an organization that relied upon community residents, rather than professional social workers from outside, to teach and train one another. MFY also invested resources in community action programs. Through its legal services unit and its community action workers, MFY counseled residents on issues of direct relevance to their lives. Staff members also advised citizens on recourse in effecting change. Lawyers, for example, informed tenants of their rights regarding landlords, including the provision to strike and refuse to pay rent when a landlord did not provide the minimum services required by law. "Inclusion of the poor," Cloward insisted, "will help to overcome a long-standing colonialism in the social welfare field."[53] The federal War on Poverty embraced this grassroots approach, mandating that federal antipoverty funds be distributed to agencies in which the poor had an active voice and local control. Tensions mounted between local community action groups and established social service agencies. The agencies feared that federal dollars were being used to harass city agencies and redistribute power within municipal government and society at large. Shortly after the Harlem riot, City Council president Paul Screvane accused MFY of instigating racial unrest through subversive tactics, including rent strikes, civil rights protests, and the printing of inflammatory literature.[54]

Community empowerment was an issue, not of "administrative efficiency and experience, but of justice."[55] As community residents gained a voice in and control over the places in which they lived, worked, and played, the political economy of inequality shaped protest and action. Health became a political watchword and rallying cry. Harlem Area Youth Opportunities Unlimited (HARYOU) was an MFY analog created in 1962 with the aid of the Harlem Neighborhoods Association and the President's Committee on Juvenile Delinquency. It saw the "problems of the Central Harlem community [as] but symptoms of a wider social pathology." Social action and social protest offered the means by which the community could "move from disorganization and pathology to health." High rates of infant mortality, tuberculosis, venereal disease, suicide, and asthma were but indicators of larger forces that suffocated the life of a community and its residents. "Nothing short of a concerted and massive attack upon the social, political, economic, and cultural roots of this pathology is required," argued HARYOU board members, "if anything more than daubing or a displacement of the symptoms is to be achieved."[56]

Housing offered a politically potent issue around which citizens and activists organized to address the integrated problems of economic, environmental, and social injustice. To Harlem residents, housing was the most visible environmental problem they confronted on a daily basis. It starkly revealed the system of economic dependency and exploitation that characterized the socioeconomic conditions in Harlem. It also highlighted the costs of a degraded environment on physical, psychological, and social well-being. Through HARYOU, Harlem youth, who might have otherwise been immersed in the politics of street gangs, got a taste of how protest and social action could effect environmental change when they worked as aides to the Community Council on Housing in one of the most sweeping rent strikes in Harlem's history.[57]

In the fall of 1963, the prominent black writer and civil rights activist James Baldwin had asked a crowd gathered in Foley Square what effect a rent boycott by Harlem residents would have on the "white economic power structure."[58] Two months later residents put his rhetorical question into action. On 1 December, with the support of the Community Council on Housing and energetic youth mobilized by HARYOU, 585 families in 50 tenement buidings in the neighborhood of 117th and 118th Streets went on a collective strike. Tenants refused to pay rent until landlords corrected countless housing violations that threatened the health and safety of residents—garbage-strewn hallways, inadequate heat, falling plaster, walls littered with rat holes, broken toilets and windows, leaky plumbing and roofs, and rats the size of small dogs. Within a matter of weeks, the strike had grown to 167 buildings and over 2,000 residents. An estimated $60,000 in rent did not flow out of Harlem into the New York City bank accounts of white landlords in January 1964.[59]

Feeling the economic pinch, Mayor Robert Wagner announced a conciliatory measure that did little to appease irate tenement dwellers. The city would launch a $1 million extermination campaign to rid the slums of rats. Rats were a public health threat. More than seven hundred people, many of them infants, had been bitten by rats in New York City the previous year. But rats were not the real public health threat. Instead, their presence, like the presence of cockroaches, revealed the larger structural inequalities that shaped the spaces in which poor communities of color lived in the United States. Wagner's proposal was like putting a Band-Aid on an internally hemorrhaging patient.[60]

In the cold weeks of early February 1964, the strike spread beyond Harlem to include buildings in the Lower East Side, the Bronx, and the Bedford-Stuyvesant section of Brooklyn. Police increasingly came to the aid of white landlords by evicting striking tenants, even though the law permitted rent boycotts when landlords failed to comply with city health and housing regulations. Many police seemed ignorant or defiant of the statute. As Jesse Gray, head of the Community Council on Housing, noted, the police reacted "with great speed to uphold the law for slumlords," but they failed to "show the same speed in arresting the slumlords and protecting the people."[61] When the police battered down apartment doors with crowbars, Gray and his staff came to the defense of striking tenants, putting their bodies on the line, as well as offering legal counsel. Gray also organized a rat-mailing campaign. Two hundred rubber rats arrived in Governor Nelson Rockefeller's office in mid-February, each accompanied by a letter signed by a Harlem resident that alerted the governor to the "faulty plumbing, unsafe wiring, lack of heat, no hot water, [and] no janitorial services" that were a "serious threat to the health of me and my family."[62] One month later, Malcolm X singled out Jesse Gray in a crowd of one thousand gathered at Rockland Palace in Harlem as a brother-in-arms in the black nationalist movement.[63] The fight for adequate housing by Harlem residents was part of a larger battle to win political and economic control of their own community. It was also an act of civil disobedience in the name of health and the environment.

In East Harlem too housing and health were at the center of a battle waged by the Young Lords, young Puerto Rican activists, for the "self-determination for all Latinos" and the "community control" of their "institutions and land." Inspired by the Black Power movement, the Young Lords, with chapters in Chicago and New York City, formed a powerful political organization in the late 1960s that agitated for social justice and community empowerment in the urban barrios of America. Fighting against "attacks" on their "land by urban removal, highway destruction, universities and corporations" and the "violence of hungry children, illiterate adults [and] diseased old people"—which they termed "the violence of poverty and profits"—the Lords launched a "garbage offensive" in July 1969. Their goal was to bring visibility to the miasma of want and disease that emanated from the streets of East Harlem. Armed not with guns but with brooms, Puerto Rican youth took to the streets to clean their neighborhood and barricade the main thoroughfares of El Barrio—Madison, Lexington, and Third Avenues—with garbage that the city's sanitation department had failed to collect. The mayor's office got the message. But the garbage carried away by the city's trucks did not remove conditions of despair.[64]

By the fall of 1969, the Lords had extended their political base and community reach through free clothing drives, a day care center, a breakfast program for school-aged children, and a free health clinic. But the Lords lacked an affordable space in which to house such programs. After a minister (a Cuban exile fearing another socialist revolution in his midst), refused to allow the use of his Methodist church as a community center during weekdays (when the building was unoccupied), the Lords and their supporters took control of the church on 28 December after Sunday services. The First Spanish Methodist Church, located in the heart of the Barrio, became La Iglesia de la Gente, the People's Church, serving the needs of over three thousand community residents for eleven days before helmeted police officers surrounded the Lords' sanctuary and forcibly removed and arrested more than one hundred people.[65]

Undeterred by the arrests and eviction, the Lords continued their efforts in community activism. Tuberculosis and lead poisoning became a focus of political action. These "diseases of oppression" plagued the people of East Harlem and were seen as symptomatic of the dilapidated housing and overcrowded conditions in which New York City's Puerto Rican population lived. On Saturdays, the Lords devoted their time to giving tuberculin tests in the Barrio. In one instance, they stole a city TB truck and screened more than one thousand residents in a single day. They also tested children for lead poisoning, using the results to prod health authorities to take action regarding the environmental risks posed to children growing up in the old tenements of East Harlem.[66]

In the politicization of health and housing, youth groups such as HARYOU and the Young Lords made visible the environmental and health disparities faced by people of color living in urban ghettos. Asthma was one such disparity. The struggle to breathe was just one symptom of a widespread pathology of urban decay grounded in economic, social, and racial inequalities. But asthma, newly visible in the urban ghetto in the 1960s, had yet to gain the political traction that more visible serious diseases, such as tuberculosis, had in mobilizing community activists. During the 1980s, rates of asthma in the inner city again climbed and once again drew media and medical attention to the disease. In this new political climate, asthma became the signature disease around which activist groups rallied to make visible the disparate environments in which Americans lived, worked, and played.

In the 1970s, the news media, physicians, and public health experts, who had first drawn public attention to asthma as an emerging inner-city disease, became relatively silent about the epidemic. By then, the spikes in asthma admissions in emergency rooms, seen in inner-city hospitals during the 1960s, had become less prominent. John Salvaggio, the physician at the Louisiana State University School of Medicine and Charity Hospital who had spent more than a decade trying to discover the cause of New Orleans's fifth season, tracing it unsuccessfully to burning dumps, grain elevators, and pollen loads, looked back on the period less through the eyes of a biomedical researcher and more through the lens of ecology and political economy. Poverty, substandard housing, and poor quality medical care combined to create a group of people ecologically vulnerable to the onslaught of allergens in their environment. The noticeable decline in emergency room admissions in the early 1970s was not due to the disappearance of any one environmental allergen. Rather, many factors combined to create a breathing space in the inner city more like that experienced by white, middle-class Americans, for whom asthma was a less serious disease—the replacement of housing stock in the neighborhood of Charity Hospital; the creation and implementation of Medicaid, which enabled indigent asthmatics to seek out private medical care; the establishment of an outpatient allergy and asthma clinic at Charity Hospital; and the availability of a new generation of asthma drugs.[67]

But the conditions that fostered a more equitable environment in the urban ghetto were short-lived. During the 1970s, the programs put in place by the War on Poverty were gradually dismantled. Between 1974 and 1986, children's poverty increased at an alarming rate, particularly in the inner cities, where 44 percent of black children lived below the poverty line. The Reagan administration's war on welfare in the early 1980s slashed new housing starts for low-income residents; dismantled Community Development Block Grants, which provided funds for housing rehabilitation; and cut federal subsidies to maintain and operate public housing.[68] At a New York Academy of Medicine workshop on housing and health in 1989, epidemiologist Rodrick Wallace and systems biologist Deborah Wallace argued that the "withdrawing of municipal services from poor neighborhoods, the resulting outbreaks of contagious urban decay," and the loss of community networks was "having profoundly serious impacts on public health and welfare." They based their statement on observed patterns of tuberculosis, homicide, elderly and infant mortality, and other indices of health and well-being.[69]

These effects had become particularly acute in public housing projects erected by the federal government after World War II. While groups like HARYOU pushed for active participation by neighborhood residents in the planning, repair, and rehabilitation of housing, federal housing policy had evolved in a different direction than the goals of community empowerment set forth by the War on Poverty. Under the Housing Act of 1949, Congress had established a federal program for the construction of public housing targeted at the urban poor. Over the next two decades, however, bulldozers demolished more low-income housing in the name of urban renewal than was built. The old five- and six-story tenement buildings came tumbling down as local authorites invoked the right of eminent domain. But little was done to relocate displaced tenants into affordable housing.[70] In Harlem, two thousand new public or publicly assisted housing units were added each year between 1949 and 1970. At the same time, demolition or abandonment by private owners took three thousand units off the rental market. By 1970, 20 percent of Harlem's total housing inventory consisted of public or publicly assisted high-rise buildings. In East Harlem, home to the greatest number of public housing projects in New York City, one out of three residents in 1965 lived in government housing.[71]

The vast complex of stark towers that came to dominate the skylines of poor urban communities in Chicago, Detroit, New York City, and other American metropolises initially provided residents with central heating, sanitary housing, and less crowded living quarters. But they also transformed once vibrant neighborhoods into concrete islands. The effect was to sever community bonds that had once been forged through daily interactions among friends and neighbors: gossiping on the front steps, shopping at the corner grocery store, watching children play on the streets. Most public housing, because of strict federal limits on building costs, was also poorly designed and shoddily built. As federal subsidies for maintenance and operation costs declined, housing authorities found themselves financially unable to address the deteriorating physical conditions that—again—threatened the safety and health of residents. In these vertical cities—building complexes that might house upwards of ten thousand people—spires of concrete concentrated and magnified the physical, psychological, and social problems they had been designed to alleviate. Malfunctioning elevators that might stop on only a few floors, faulty heating and ventilation systems, leaky plumbing, infestations of rats and cockroaches, and high rates of crime and drug addiction were just a few of the problems that began to surface in the late 1970s and through the 1980s.[72] Another was asthma.

By the later 1980s, after nearly two decades of relative quiet, public health officials, physicians, the popular press, and community activists again turned the national spotlight on an emerging public health crisis in America's inner cities. An initial wake-up call came when the Centers for Disease Control reported in the late 1980s that the number of individuals with asthma had increased 29 percent nationwide between 1980 and 1987. In 1998, the statistics were even more alarming: the prevalence of asthma between 1980 and 1996 had increased 75 percent.[73] Although changes in diagnosis might account for some of the observed increase, most medical and public health professionals saw the trend as a real effect and not as an artifact of changing classification categories. Mortality and morbidity patterns revealed widespread geographic variation. The prevalence and severity of asthma appeared especially acute among inner-city African American and Hispanic populations. Between 1982 and 1986, for example, hospitalization rates for asthma in East Harlem were sixteen times higher than those in Greenwich Village–Soho.[74] Minority children seemed especially vulnerable. Compared to whites, African American children were almost twice as likely to suffer from asthma and 2–5 times more likely to die of the disease.[75] In Central Harlem, a 2005 study found that one in four children had asthma. The national average was one in sixteen.[76] Suddenly the nation found itself confronting a "new" urban asthma epidemic, while sadly remaining oblivious to history and the lessons of the first epidemic outbreaks in the 1960s.

Medical opinion has once again divided over whether poverty or race accounts for the observed disparities in asthma.[77] Medical and public health researchers debate whether accessibility to health care, adverse effects of asthma medications, environmental exposures, or different genetic susceptibilities can explain the current trend. At the same time, these professionals acknowledge that no single factor is likely to offer an adequate explanation of the urban asthma epidemic, even as medical research largely views urban asthma through the lens of individual behavior and pathology. Large, federally funded research projects like the National Cooperative Inner-City Asthma Study examine high levels of exposure to cockroach allergen, tobacco smoke, and other indoor materials in hopes of understanding and controlling inner-city asthma. Other studies, such as the decade-long Collaborative Study on the Genetics of Asthma, funded by the National Heart, Lung, and Blood Institute, search for underlying genetic differences both within and across African American, Hispanic, and Caucasian populations. Such studies aim to unmask the genetic factors involved in asthma—for example, different genetic susceptibilities to cockroach allergens—in the hopes that they will lead to new treatment therapies targeted at the molecular level. Despite different emphases on environment or heredity, these multimillion dollar research efforts frame the problem of urban asthma largely around individuals and their ability to manage and control the disease and the spaces in which they live.[78] But is asthma one person's disease, or is it society's problem?

Community activists who live and work in inner-city asthma zones confront on a daily basis the burden the disease places on children. Such activists tend to see asthma in ways different from the views of the biomedical community. "You can't just get rid of cockroaches and expect asthma to go away," remarked one organizer of West Harlem Environmental Action (WE ACT). "For that matter, you can't just put in better buses and expect asthma to go away. It's all got to be approached in a social justice framework."[79] A puff of an inhaler might ease the symptoms of asthma, but it will not alleviate the underlying environmental and social conditions that exacerbate the disease.

Founded by Peggy Shepard and Vernice Miller-Travis in 1988, WE ACT is a nonprofit, community-based environmental justice group built on the principles of self-determination and community empowerment to "fight environmental racism and improve environmental health, protection and policy in communities of color."[80] Shepard, a journalist by training, got her first taste of social activism working for New York State's Division of Housing and Community Renewal in the late 1970s. Over the next decade, she and others watched as people in poor communities of color in Warren County, North Carolina, and in Houston (among others) drew upon protest strategies and legal actions learned during the civil rights era to speak out. These communities challenged in the courts the disproportionate share of hazardous waste facilities, garbage dumps, and polluting industries that ended up in "poor, powerless, black and Latino communities, rather than in affluent, white suburbs."[81] A new social movement—environmental justice—gathered political momentum in the 1980s. But its roots reached into the past, into the era of civil rights and the take-it-to-the-streets fight for health and social justice instigated by groups like the Young Lords.

When New York City built a sewage treatment plant on the North River at 137th Street, after an unsuccessful bid to locate it along the Hudson River at 70th Street in upscale Manhattan, Harlem residents were outraged at the foul odors that emanated from the poorly managed facility. Besides the stench, many residents experienced difficulties breathing after the plant opened in April 1986. People took action. The "Sewage Seven," which included Shepard and six others, blocked traffic on the West Side Highway in front of the plant on Martin Luther King Day in 1988. Other community residents, donning gas masks and carrying placards, held up traffic on Riverside Drive. After a four-year legal battle and continued community mobilization, Mayor David Dinkins earmarked $50 million for cleanup of the plant. The legal settlement also required the city to pay $1.1 million to establish a West Harlem fund to address the community's environmental and health concerns.[82]

The North River sewage treatment plant is just one of many environmental burdens faced by Harlem residents. Six out of eight Manhattan bus depots are located in northern Manhattan. Major commercial truck routes for moving goods in and out of Manhattan also pass through the center of Harlem. Because Harlem sits in a valley, the concentration of sulfur dioxide and particulate matter emitted in diesel bus and truck exhaust is intensified. After WE ACT's repeated efforts failed to get the Metropolitan Transit Authority (MTA) to update city buses so that they could use cleaner fuels, the organization filed suit against the MTA in 2000 with the U.S. Department of Transportation. WE ACT charged that the MTA had violated the "civil rights of Northern Manhattan's mostly black and Latino residents" by disproportionately locating diesel bus depots in minority communities. "Given the large number of Harlem children with asthma who are vulnerable to pollution-related asthma attacks," said WE ACT's environmental health director Swati Prakash, "it is more important than ever for our leaders to step up to reduce the disproportionate burden of diesel exhaust and other sources of air pollution in our communities."[83]

Asthma is the disease at the center of WE ACT's ongoing fight against economic, social, and environmental injustice. Asthma is, as Brown University sociologist Phil Brown wrote, a "stepping point to a politicized view of the world."[84] Environmental justice groups like WE ACT, through partnerships with Columbia University's Mailman School of Public Health, have developed community-based research programs. In these programs, local citizens and youth play active roles in the investigation of harmful exposures in their neighborhoods and in the education of residents about individual and collective actions to protect their communities from environmental harm. Youth from WE ACT, for example, became environmental sentinels. By wearing personal air monitors that can measure diesel exhaust particulates, polycyclic aromatic hydrocarbons, pesticides, and other hazardous substances found in their homes and schools and on their street corners and other neighborhood hangouts, these youth helped to reveal the places and kinds of exposures that most jeopardized their health.[85] Similarly, in the Greenpoint-Williamsburg neighborhood in Brooklyn, where asthma rates are twice the national average, El Puente, a community learning and development organization, has enlisted the support of high school students and volunteers in community health research. El Puente's project is modeled after the tradition of the Young Lords' "barefoot doctors," who raised awareness of the health and housing conditions in East Harlem through their focus on lead poisoning. The Greenpoint-Williamsburg neighborhood is a poor, ethnically mixed community of Latinos, African Americans, and Hasidic Jews. It is also home to a sewage treatment plant, thirty solid waste transfer stations, thirty hazardous waste storage facilities, and seventeen petroleum and natural gas storage areas. Local residents, through their intimate knowledge of place, made visible the harmful environmental exposures that outside researchers had failed to notice or consider. When, for example, El Puente unexpectedly found high rates of asthma among women over the age of forty-five, focus group discussions revealed that many of them worked in laundries, dry cleaning establishments, hair and nail salons, and textile factories—all occupations, notes Columbia University urban planner Jason Corburn, where the risks of respiratory disease are great.[86]

Through community-based participatory research, El Puente executive director Luis Garden-Acosta states, science becomes "an instrument for collective self-help."[87] Investing in the gathering of information about the environment in which they live, work, and play empowers citizens to act. Through the research and educational efforts of groups like WE ACT and El Puente, people suffering from asthma in the inner city have begun to see themselves "less as an individually sick person" and more as a collective of people living in a world where exposures, both indoor and outdoor, are a consequence of an urban ecology—physical, economic, and social—shaped by a long history of racism and economic inequality in America.[88]

It is almost a half century since Flavio da Silva stepped off the plane in Denver, brought here to be cured because Americans took notice. It is also almost a half century since the first wave of urban asthma epidemics swept the United States. Since that fateful time that forever altered Flavio's life, three generations of children in America have grown up in conditions of poverty, poor housing, and despair. How many epidemics of asthma does it take to notice the ecology of injustice that makes each breath a fight to survive?

CHAPTER

5

On the Home Front

I don't care about pollution
I'm an air-conditioned gypsy
That's my solution.

—The Who, "Going Mobile," 1971

When Charles Davies graduated from the School of Engineering at the University of Buenos Aires in Argentina in 1917, the indoor environment offered a new frontier of research and development for a freshly minted engineer. Technological advances in heating, cooling, and ventilation promised modern comfort in the home and office. Enterprising firms like Carrier Engineering Corporation were selling man-made weather for factories, office buildings, movie theaters, and department stores as America moved into the prosperous 1920s. So Davies set off for New York City, where skyscrapers soared into the sky and became symbols of the technological prowess and bold confidence of engineers in mastering the great indoors. At the newly opened Madison Square Garden, itself an engineering triumph, the first National Heating and Ventilating Exposition, held in 1926, offered a showcase of the latest products—furnaces, refrigerators, air-purifying and air-cooling systems—that promised Americans relative freedom from nature and its seasons. It must have been a dazzling sight to a young man who had staked his future on this new realm of engineering.[1]

But other New York City sights caught Davies's eye—in particular an intelligent young woman with bobbed hair named Isabel Beck. In 1923,

Beck became one of the first women to graduate from Columbia University's College of Physicians and Surgeons. She, with a passion matching Davies's, had great faith in the powers of science and medicine to engineer humanity's progress. When Beck met Davies in the mid-1920s, she was the only female intern at Mount Sinai Hospital. But neither her struggles as a woman for an equal footing in medicine nor her marriage to Davies deterred her strong sense of economic independence and reproductive freedom. These ideals shaped the career choices she made both as a female physician and as founder of Planned Parenthood of Southern Westchester. Beck and Davies shared a commitment to a modern marriage and home, guided by advances in science, technology, and medicine.[2]

Despite his professional prospects and whirlwind romance, not everything about modern American life agreed with Davies. During his first spring in New York, he suffered from miserable cold-like symptoms—tearing eyes, violent sneezing, and gasping breath—that did not let up as the season progressed. Beck, a resident at Mount Sinai Hospital's recently established Children's Asthma Clinic, suspected that the timing of her husband's sneezing fits had little to do with their recent marriage and everything to do with the pollen of North American grasses. Fearful that the sexual exploits of trees and grasses would put a damper on the couple's honeymoon bliss, Beck dragged her husband to the clinic, where she subjected him to the latest tools of the allergy trade—scratch tests and desensitization treatments—in the hopes of alleviating his symptoms. But the shots brought no relief. Davies was one of an estimated 35 percent of allergy sufferers who failed to respond favorably to immunotherapy.

Undeterred by the clinic's failure to give Davies relief, the couple combined their talents in engineering and medicine to build a home environment that imitated the natural environment in hay fever resorts: a pollen-free atmosphere. The idea came to Davies when he stumbled into one of New York City's newly air-conditioned motion picture palaces and found refuge from his illness. Theaters, such as the Rivoli near Times Square, lured patrons by promoting their manufactured climate of "refrigerated mountain ozone" during the sweltering days of summer.[3] Motion picture theaters were the first spaces where the majority of Americans first encountered a man-made climate. Although air-conditioning was beyond the reach of all but the wealthiest homeowners in the 1920s, Davies knew from his wife's medical expertise that the comfort he found inside the theater most likely came from the absence of pollen, not the refrigerated air. Creating the ideal indoor climate entailed far more than temperature alone. Refrigerated air-washing systems designed by engineering firms cooled the air but also cleansed it of dirt and dust. Davies believed that mechanical filtration, a beneficial by-product of air-conditioning, could be adapted readily to the home environment at a cost the average hay fever sufferer could afford. It was a serendipitous discovery that led the couple into an enterprising business partnership.

As the couple tested various filter materials and fans that would keep out pollen while allowing enough air movement to keep their apartment comfortable on a hot summer night, Davies's sneezing served as a "barometer" of their "lack of success." Daunted by repeated failures, Davies briefly worked on a gas-mask filter for street wear before hitting upon a promising cellulose filter. The filter, combined with a motor and fan, successfully eliminated a fifteen-by-twelve-foot room of dust and pollen and exchanged the air every five to twenty minutes. Eleven hours a day spent in this pollen-free room and Davies was ready to brave the streets of New York. If it worked for him, it might also benefit millions of other weary sneezers who could not afford the time or expense of a hay fever holiday. To find out, the couple invited hay fever victims to sleep in their pollen-free apartment on West 111th Street. Optimistic reports by patient-guests prompted the couple to take a patent out on their window filtration unit. Beck decided next to subject their invention to a more rigorous clinical trial.[4]

With the aid of her colleague Murray Peshkin (the parentectomy doctor), at the Children's Asthma Clinic at Mount Sinai Hospital, Beck recruited fifty-four patients for the study. She chose people whose hay fever or asthma had improved in previous years at hay fever resorts, on ocean voyages, or during residence abroad. The mechanical filtration device was installed in patients' homes. In addition, the doctors prescribed other domestic alterations and regimens meant to free the indoor environment of offending allergens. Mattresses were to be vacuumed and covered with a thick cotton slip. Cotton or air pillows were to be used in place of feather pillows. Dry dusting or sweeping was to be avoided. So too was any contact with talcum powder or the family pet.

The Airgard-Equipped Room is a Miniature Hay-Fever Resort at Home

Causative pollens removed from the air by the Airgard

Specialists in allergic diseases have found that the Airgard, when used in connection with the routine pollen serum treatments, is very often an aid in building up immunity to hay-fever. Many patients who spend from eight to ten hours in the pure, pollen-free atmosphere provided by the Airgard find that their systems recuperate sufficiently to allow them to go about freely the remainder of the day without ill effects.

The Airgard is a small, portable cabinet, so low in cost that its benefits are available to everyone. It fits easily and quickly into any sliding window. Two models are offered: Model "B", which is for general air cleaning in offices and homes; Model "C", which has twice the capacity in the same size cabinet and is equipped with variable speed motors and switches, for use in connection with treatment of allergic diseases.

Send for Free Booklet

Write for free literature giving prices and full information. Or mail this advertisement, with your name and address, to American Air Filter Co., *Incorporated,* 131 Central Ave., Louisville, Ky.

AMERICAN AIR FILTERS

Figure 35. Airgard turned any room into a "miniature hay-fever resort at home." Courtesy of American Air Filters International.

Within forty-eight hours of their voluntary isolation from the outdoors, the majority of hay fever and asthma patients reported adequate relief.[5]

As word of their invention quickly spread, Beck and Davies were inundated with orders from hospitals, laboratories, and allergy clinics and were able to raise enough capital to manufacture their filter on a large scale. By the early 1930s, manufacturing firms began advertising their own home pollen-filter units. American Air Filters claimed that its product, Airgard, was able to turn any room into a "miniature hay-fever resort at home." Airgard was "so low in cost" that its benefits were "available to everyone"; "everyone" meant the middle-class American consumer who could afford home comforts beyond the immediate necessities of shelter, food, and clothing (Figure 35).[6]

The retreat to the great indoors in search of allergy relief had begun. Engineers and physicians believed they could build safe havens in the homes of allergy sufferers that were far better than nature afforded in even the most popular of hay fever resorts. This belief was reinforced by the technological optimism of a nation and the confidence of heating, ventilation, and air-conditioning engineers in their ability to control the indoor environment. But the desire to be free of nature through the creation of an artificial environment was an unrealistic quest. The home, it would be discovered, had its own ecology, shaped by changing relationships of building materials and design, home furnishings and décor, and behavior patterns of human and nonhuman residents. It was an ecology not so easily ignored. In the historical landscape of the home, different allergens, different understandings, and different solutions to the problem of allergic illness came into being. How Americans have built, thought about, and experienced the indoor environment where they have lived and worked over the last century is a story in which allergy plays a part.

The partnership forged between Isabel Beck and Charlie Davies was just one of many such collaborations between physicians and industrial engineers during the interwar years as these innovators worked to build a breathing space for allergy sufferers in the comfort of home. Cooperation between allergists and botanists brought out aspects of the local natural environment that informed therapeutic practice. In the same way, alliances between doctors and engineers, and between doctors and their patients, brought into view a complex picture of the indoor environment.

Indoor allergens and their control increasingly became a focus of research, the development of technological solutions, and public concern.

In moving to the study of the indoor environment, allergists found themselves adopting ideas and methods that had first been worked out on the factory floor. Milton Cohen, a Cleveland physician, first became aware of the promise engineering offered to his medical specialty when a middle-aged man came to his allergy clinic in January 1924 seeking relief from seasonal hay fever and asthma that would soon be sure to afflict him. For almost eighteen years, the gentleman had been a consulting engineer to the dusty trades—paint factories, textile mills, foundries, and mines—but he was so incapacitated during the summer and fall that he either fled to the country or suffered at home in the city, unable to work. In the course of their conversation, Cohen learned from his patient that with the right equipment, know-how, and capital, industrial plants could keep conditions on the factory floor relatively free from fine dusts. Industrial America had increasingly become a world of exposures. And dusts, such as those from lead and zinc oxide, had become of increasing concern to public health experts, social reformers, insurance companies, factory workers, and labor unions in early twentieth-century efforts to improve the factory environment, social conditions, and health of America's workers.[7]

But dust could be made of many things. When Cohen informed his patient that his nemesis, pollen, was also a form of dust, the engineer applied his expertise in environmental control on the shop floor to the home. During an August business trip to Detroit, the patient was overcome by a severe attack of asthma, one that even hourly injections of epinephrine could not control. The patient retreated to his bedroom and turned on the prototype pollen filter he had designed. Within three days, his attacks had ceased. As long as he spent two-thirds of each day in a pollen-free atmosphere, the engineer was able to weather the outdoors symptom-free, even at the height of the pollen season.[8]

Experimental trials of the pollen filter, conducted by various universities, hospitals, and allergy clinics, confirmed what the engineer's own body had told him. The device, which looked like a barbecue grill made out of a metal drum, filtered out 99 percent of pollens and 62 percent of tobacco smoke in the air. Lined with a filter made of woven wool and cotton cloth and outfitted with a variable-speed fan capable of delivering two hundred cubic feet per minute of fresh air, the invention created a slight indoor positive pressure that forced old air out of cracks and crevices, eliminating the need to tightly seal the room. Within a short time, Pollenair hit the commercial market. For $150—the price of a new Victrola Granada phonograph—hay fever sufferers could enjoy air emanating from an elegant nightstand "as pure, as clean, as the ozone . . . in the big North Words or on the Mountain Tops."[9]

To the hay fever victim, a pollen-free room offered respite from irritation and suffering. To the clinician, it offered an experimental laboratory. Pollen was the first and most common allergen to be identified as a causal factor in the onset of hay fever and asthma. Yet other exposures also appeared to exacerbate symptoms. House dust, pet dander, cosmetics, and various foods were among the growing list of likely suspects that allergists identified as potential allergens.

To make visible the threat of allergens that lurked in the indoor environment, allergists turned once again to engineers. By the 1920s, a new science of industrial hygiene was emerging. In programs at Harvard and Yale and in federal agencies such as the U.S. Bureau of Mines, doctors and engineers had begun to replace social workers and union activists as the investigators of the environment, health, and comfort of those who toiled in the factories, mines, and office buildings of modern America. The site of their research was the laboratory. In a hermetically sealed room with a human subject, engineers could isolate individual environmental factors—carbon monoxide, temperature, or humidity, for example—and measure the physiological effects on the human body. These so-called psychrometric chambers were common in the early 1920s in places like Harvard's Department of Industrial Hygiene and the U.S. Bureau of Mines' American Society for Heating and Ventilation Engineers (ASHVE) Research Laboratory. Allergists readily employed the engineers' designs of the environmental chamber in building a pollen-free room. They also embraced a mechanistic model that sought to isolate single cause-effect relationships between a particular exposure and the body's response.[10]

Simon Leopold and his younger brother Charles were the first to design and build a modified psychrometric chamber for the study of the allergenic properties of the indoor environment. Simon was a graduate and faculty member of the University of Pennsylvania's School of Medicine. Charles was one of the nation's most sought-after consulting engineers in the design of air-conditioning systems. He had brought cool comfort to Madison Square Garden, the New York Stock Exchange, the Capitol in Washington, D.C., and more than sixty Warner Brothers Theaters during the 1920s. Together in the early 1920s the Leopold brothers built a combination laboratory-hospital room at the University of Pennsylvania in Philadelphia; the room had plastered walls; a waxed, wooden floor; and tightly sealed windows and doors. Through the use of an air washer, fan, and blocked ice, they could precisely control the temperature, humidity, and dust concentration in the room's atmosphere. Allergists didn't know which climatic factors other than pollen influenced the health and comfort of asthmatic patients or how they did so. By employing an engineer's precision, Simon and Charles Leopold hoped to find out.[11]

Poor Mr. G. A fifty-one-year-old bronchial asthmatic, he had been suffering daily from severe asthma attacks for almost five years when he came to the University of Pennsylvania hospital in January 1924. Unable to work, he spent most of his days in Fairmount Park. Built in 1855, the park was created to protect the health and water supply of Philadelphia's citizens. In open air along the banks of the Schuylkill River, Mr. G found temporary relief from his wheezing attacks, which were provoked by confinement in a stuffy room. A battery of scratch tests showed no sensitivity to foods, pollens, or animal dander, with the exception of that of cattle. In the hospital, Mr. G would spend his days and nights by an open window. Epinephrine injections were his only hope for a decent night's rest. The Leopolds couldn't have hoped for a more perfect experimental subject.

After a week in the hospital, Mr. G was escorted into a new private room. Except for the absence of other patients, to him it appeared like any other place in the hospital ward. But the air washer system had been hard at work cleansing and cooling the air during the previous day. After he had spent twenty-four hours in his new room, Mr. G's symptoms miraculously disappeared. For the first time in ages, he was able, without epinephrine injections, to sleep fully reclined, with only a single pillow under his head.

Five days later, the Leopolds had Mr. G move out of the special room and back into the regular ward. Within two hours, Mr. G was gasping for breath and begging to return to the special room. The staff, under orders, refused his request. At midnight, an intern on duty saw the man in great distress and permitted him to return to the environmental chamber. In less than an hour, Mr. G was sleeping comfortably in his old bed.

Something or a combination of things—temperature, humidity, dust-free air—gave Mr. G relief in the isolated room. The Leopolds had successfully eliminated every potentially offending substance in Mr. G's environment and produced relief. Now, unbeknown to Mr. G, they began subtly to modify his surroundings and observe the effects. With a vacuum cleaner as their scientific instrument, the Leopolds collected two ounces of finely powdered house dust from another location. When Mr. G left his room for a few minutes, the brothers secretly scattered the house dust across the floor, taking care to avoid any visible clumps of dust. Mr. G returned, and within forty-five minutes he screamed that he was suffocating and felt like he was about to die. Attendants rushed him out in a wheelchair, placed him by an open window, and shot him up with epinephrine until his attack subsided. It was a close call. The Leopolds' experiment was one of the earliest dramatic demonstrations of house dust as a potent indoor allergen. It also showed the promise of engineering in allergy relief.

The air-conditioning industry and the clinical field of allergy both profited from the partnership between engineering and medicine. In their efforts to manufacture an ideal indoor climate, engineers looked to the places that attracted hay fever sufferers during certain seasons—mountain spas and beach resorts—for clues into the relationships between climate and comfort. But to imitate nature was no simple task. To achieve indoors what nature accomplished at its very best required the right combination of temperature, humidity, and air circulation. As engineer Frank Hartman explained: "There are many synthetic products of the laboratory that are superior to those made in nature's workshop, but when we resort to the artificial to replace nature, we must be sure to include all that nature provided, or our artificially created conditions will fail us."[12] Nature could be unpredictable as well. Wind direction and speed might instantly change, humidity and temperature levels might drop or rise, and even the most pleasant vacation at the seashore or on a mountaintop might turn into a miserable stay. Hay feverites frequenting hay fever resorts knew this only too well. An artificial climate that delivered constant comfort 365 days a year offered a decided advantage over a healthful but fickle nature. Consequently, air-conditioning, which aimed to control temperature and humidity in addition to dust, held an edge over mechanical filtration. The difficulty was cost.

Despite the stock market crash of 1929, corporate executives remained optimistic about the continued growth of the air-conditioning industry. In less than a decade, the public had become accustomed to and even expected a cool indoor climate in office buildings, department stores, and motion picture palaces. But would Americans embrace air-conditioning as a way of life in their homes? General Electric (GE) executives believed so. In 1932, they boldly asserted that home cooling was a $5 billion market ready for the taking. Companies like GE and Frigidaire entered the cooling business with gusto. But few Americans in the 1930s rushed to buy the new $400 single-room coolers, let alone the more expensive heating, ventilation, and air-conditioning (HVAC) systems capable of controlling indoor climate throughout an entire home. In 1938, only fifty-five thousand of an estimated 22 million electrified homes had air-conditioning.[13]

The comfort of climate seemed to be a luxury most Americans could live without, particularly in an atmosphere of economic depression and war—unless, that is, they suffered from allergies. Then luxury became necessity. Four million Americans—roughly 3 percent of the population—had hay fever. And the number was on the rise, William Welker, head of the Allergy Unit at the University of Illinois College of Medicine, told an audience of heating and ventilating engineers in 1936.[14] Given the depressed market, the HVAC industry would do well to cultivate the allergy trade. With the aid of American Air Filters, GE, and Frigidaire, Welker had been conducting research on the benefits of air filtration and cooling for victims of hay fever and asthma. Frigidaire also funded the research of Leslie Gay, a leading allergist at Johns Hopkins University. Gay's task was to furnish conclusive scientific evidence of the value of air-conditioning in the treatment of hay fever and asthma.[15] In the early 1930s in the *Journal of the American Medical Association,* both Welker and Gay published studies emphasizing the advantages of air-conditioning over air filtration. Welker and Gay attributed the major source of relief to pollen filtration. But they also found that by reducing the level of relative humidity, air-conditioning offered more to improve the health and comfort of hay fever and asthma patients.[16] In furnishing an "artificial climate as nearly as possible like the refreshingly cool and pollen-free atmosphere of the Rocky Mountain and the Northern Great Lakes States," Frigidaire, GE, and other makers of room air conditioners welcomed the medical seal of approval and used it in their advertising efforts.[17] Some Americans, rich or poor, might balk at spending $750 for an appliance that did little more than take the prickly edge off a hot summer night. But for Mrs. Hanna Kellogg, who married into the fortune, fame, and family of Kellogg's Corn Flakes, it was a small price to pay. She purchased a GE room cooler in 1931 and reported a night of "restful sleep" better than any she had had in eighteen summer seasons at hay fever resorts at the mercy of nature.[18]

In the 1930s and 1940s, only a privileged few could purchase allergy relief for the home. Some of the less fortunate allergy sufferers organized their lives to take advantage of air-conditioned spaces available to the public at large. Hay feverites ferreted out such spaces and traded experiences. In Chicago, certain air-conditioned buildings became known locally as "hay fever resorts."[19] When the University of Kentucky equipped Dicker Hall, a social club for students in the College of Engineering, with a unit air conditioner, the hall quickly became a haven for local hay feverites.[20] In Lincoln, Nebraska, the Lincoln Liberty Life Insurance Company reported unforeseen benefits from its 1936 investment in modern air-conditioning while it was renovating its headquarters office building. Not only did the new environment improve worker efficiency and reduce machine breakdowns, but it also proved to be a haven for tenants and employees afflicted with hay fever.[21] Some sensitive souls even chose their occupation according to their air-conditioning needs. Katherine Madison, a social worker at an allergy clinic in one of New York City's large hospitals, wrote of how one patient with an extremely severe case of hay fever obtained a job as an usher in an air-conditioned movie theater, ate only in air-conditioned restaurants, and slept in an air-conditioned hotel room.[22] When he did venture outdoors, he wore a patented nasal air filter. This filter was like the one worn by Paul Derringer, star pitcher of the Cincinnati Reds—nicknamed the "Kentucky Rifleman"—when his hay fever threatened the team's chances at two successive pennant races and the world championship in 1940. Hannah Lees, a correspondent for *Collier's*, jokingly told readers in 1936 that with modern treatment, "seasonal sneezers can angle for jobs in air-conditioned office buildings and stores and between times run from air-conditioned bar to restaurant to movie to train to grillroom." Her description was an accurate portent of the allergic lifestyle and of modern American life.[23]

When the Houston Astrodome opened in spring 1965, the move to the great indoors was virtually complete. A nine-acre bubble-covered ballpark of artificial climate and artificial grass, the Astrodome sheltered baseball fans and players from inclement weather, buzzing mosquitoes, and pollen. Even the nation's favorite outdoor pastime could now be played inside. While engineers envisioned even bigger domes that could safely contain and insulate entire cities from external natural and man-made threats, the home in postwar America had already become an ideal, bubble-encased world. In a 1969 essay that encapsulated the technological optimism and fears of Cold War America, *House Beautiful* author John Ingersoll painted a portrait of two different homes in an American suburb. It was a place where "sulphur-streaked, gray smoke billows from stacks high above an industrial plant," jets thunder off an airport runway to the north, and "a pungent odor drifts into the neighborhood from a tannery across the river." In one home, a mother looks out on this scene through her picture window and is not alarmed. Her children are safe inside, protected from toxins, including pollen, long present in nature, and more recent additions of the nuclear age, including sulfur dioxide, radioactive fallout, and DDT.[24] Science and technology had created some of the menacing threats outside her window. But science and technology had also engineered the indoor environment in which her family was secure. Inside her house "is an atmosphere as exhilarating as a cool day with warm sun: as peaceful, clean, and sweet-smelling as a spring morning on an Iowa farm." How grateful she is to afford the latest in HVAC engineering: a "home pleasantly cool in summer, comfortably warm in winter, never too dry, and virtually dustless at all times."[25]

In the house next door, a different scene unfolds. Here "a mother wipes away from the window sill a thin layer of soot, and strains to hear the radio weather report as the jet roars overhead." Her three-year-old child, "playing on the living room floor, shivers slightly" as a cold draft blows across the room. What she would give to have her neighbor's "total comfort package"—an electronic air cleaner, central air-conditioning, a warm-air heating furnace, and a humidifier to transform her living space into a healthy, sheltered environment. She and her husband were eligible for a tax deduction after the doctor prescribed an electronic air cleaner to ease their child's allergies. But the $3,500 price tag—the same cost as the new car her husband had his eye on—gave her pause.[26]

Ingersoll's article affirmed the home as a protective retreat from the risks Americans faced or feared in the aftermath of World War II. Every day, it seemed, new dangers came to light in newspaper headlines or on radio and television news. Communism was a threat Americans had come to live with; it was a remote but ever-present white noise, hissing amid the clamor of crying babies and consumerism, which occupied American family life during the Cold War. Fears of nuclear fallout sounded more clearly. The St. Louis Baby Tooth Survey was a massive effort initiated in 1958 by the St. Louis Committee for Nuclear Information to collect over fifty thousand baby teeth from children born in the nuclear age. Its findings revealed high levels of the radioactive isotope Strontium-90 and suddenly made exposure of the American public to risky radiation levels visible and real. "A tide of chemicals [had] arisen to engulf" the environment, Rachel Carson warned Americans in her 1962 book, *Silent Spring*. Hazards quite different from disease organisms responsible for the smallpox, cholera, and plague that had once imperiled people and nations now lurked in the environment. These dangers were a consequence of modern life. Even if their presence was "formless and obscure," Carson wrote, they were "no less frightening." Caught in this ecology of fear, Americans looked to the home as a safe place, sheltered from the uncertain risks and exposures that lay hidden in the outside world.[27]

As the number of American families escalated in the 1950s, increasing 28 percent from 1947 to 1961, so did the number of single-family houses. Housing starts in 1955 reached an all-time high: 1.65 million new homes were built, more than ten times the number a decade earlier.[28] A single-family home had come within reach of the average middle-class American. The home had long been regarded as a place of refuge from a chaotic and at times dangerous world. But Americans embraced this image of the home as a haven with renewed enthusiasm in the postwar years. Manufacturers, eager to sell household appliances and goods, vigorously promoted this ideal. The home was a "self-contained universe" where all the necessities and luxuries of life—food, comfort, leisure, and entertainment—could be had. A happy home also came to include a fixed and comfortable indoor climate. Thus in the 1950s, air-conditioning manufacturers, hoping to cash in on a remarkable 240 percent increase in the sale of home appliances and furnishings, sold indoor comfort as a path to family togetherness.[29] "You enjoy your children more in a Weathermaker Home," exclaimed a 1956 Carrier ad. "We're healthier," "We sleep better," "We dust less," exclaimed proud owners of air-conditioned homes who were surveyed in Dallas and Houston by *House and Home* in 1954. Another homeowner, Herman Blum, even boasted that "he and his wife almost never use their outdoor terrace. Not only is it cooler inside but there is also an absence of mosquitoes and bugs. . . . As soon as it is warm enough for outdoor living it is time for air-conditioning." Such ads proved successful. By 1960, more than 12 percent of American homes had either a central cooling system or one or more room air conditioners. What had once been considered an extreme luxury had in just over a decade become, manufacturers told consumers, a "plain necessity." More time indoors meant more time with the family in the private, comfortable, secure environment of home.[30]

To allergy sufferers, the engineered indoor environment that secluded them from the hazards of pollen offered technological control over nature—or at least its most unpleasant aspects. Hay fever and asthma patients could successfully avoid the worst of the fall hay fever season. Yet many found that some symptoms persisted into winter. Such victims of what *Saturday Evening Post* writer Steven Spencer referred to as "ragweed hangover," through their refined powers of perception, helped doctors identify allergens besides pollen that inhabited the home. More time spent indoors also meant that new risks came into view, from which even man-made weather offered no guarantee of escape.[31]

House dust was the first notorious indoor allergen to capture medical and engineering attention. In 1918, a young soldier walked into the recently opened allergy clinic at New York Hospital complaining of asthma. The soldier was on furlough, waiting to ship out for the trenches in Europe. Something curious had happened to the recruit while he was away at boot camp in Texas: his asthma, which had plagued him since childhood, had disappeared. He was back in New York only a few days before his asthma returned. The standard battery of pollens the clinic used in skin testing gave no positive reaction. So Robert Cooke, head of the clinic and a pioneer researcher on allergic disorders, gave the soldier orders to return with a vacuum cleaner bag filled with dust gathered from his home. When the recruit returned, Cooke prepared an extract of the house dust and injected it under the patient's skin. Immediately a sizeable welt appeared on his arm, and he was overcome by an asthma attack. Cooke's findings, published in 1922, offered the medical community the first clinical proof of house dust as a potential indoor allergen. They also launched an almost fifty-year quest into the mysterious allergenic properties of the home's most ubiquitous guest.[32]

Every effort to hunt down the "insidious agent" hidden in house dust seemed to lead to a dead end.[33] House dust was not one thing. Its composition varied "in different parts of the country, in different seasons, and from house to house."[34] It came from draperies and curtains, pillows and bedding, people and pets, carpets and clothing, cosmetics and toiletries, and countless other sundries, inanimate and living, that were in the home. Every home had its own ecology of dust that defied standardization, no matter how hard manufacturers of allergenic extracts tried to standardize it (Figure 36). The best one could do, Cleveland doctor Milton Cohen observed in 1929, was to go into a patient's home with a vacuum cleaner and collect dust from the "mattress, pillows, and coverings of the patient's bed, from the living room rug, from overstuffed living room furniture, from the bedroom rug, from the automobile and from various other sources that suggest themselves after a survey of the environment."[35] Then a house-dust extract specific to the patient could be prepared for the purpose of desensitization. Mold, spores, animal dander and hair, feathers, orris root, flaxseed, cottonseed, pyrethrum, silk, and insect emanations, to mention only a few ingredients commonly found in house dust, were key suspects that could unleash an asthma attack. In 1949, the American Academy of Allergy (which changed its name to the American Academy of Allergy and Immunology [AAAI] in 1982) instituted and funded a "dust fellowship" to address what allergists of the day regarded as the "major problem in the field of inhalant allergy." But even scientists at the California Institute of Technology could not uncover, through the most advanced biochemical methods available, the nature of the house-dust allergen. After five years, the American Academy's Subcommittee on Dust abandoned the fellowship, leaving further work on the immunologic activity of house dust to individual researchers.[36]

Lack of knowledge regarding the biochemistry of house dust did not hinder the development of preventative measures designed to alleviate the suffering of allergic patients inside their homes. Responsibility fell

From Maine to Florida . . . from Washington to the Gulf ports, patients allergic to house dust can now be treated with a single preparation. No longer is geography a factor in diagnosis or treatment of house dust allergy. Nor is the matter one of variation in local dusts. ALLERGENIC EXTRACT PURIFIED HOUSE DUST CONCENTRATE – ENDO is prepared by an original process* which eliminates these considerations.

Enjoying a record of over 90% diagnostic accuracy, this unique product also exhibits a marked degree of uniformity . . . is effective with the scratch-test method and, thus, simple to employ.

Wherever you are testing or treating for house dust allergy, you can rely upon the universality of

ALLERGENIC EXTRACT
PURIFIED HOUSE DUST CONCENTRATE-ENDO

Write for complete literature.

Diagnostic Package: 0.5% (1/200) solution Allergenic Extract Purified House Dust Concentrate Diagnostic in glycerosaline solution, 1 c.c. applicator vials.

ACCEPTED
AMERICAN MEDICAL ASSN.

Treatment Set Package: Allergenic Extract House Dust Concentrate Therapeutic in graduated concentrations from 1: 40,000 to 1:40. Also available in bulk package, 1:40 concentration.

*Prepared, under exclusive license, by Boatner-Efron process, U. S. Patent No. 2,316,311.

ENDO PRODUCTS INC. RICHMOND HILL 18, NEW YORK

Figure 36. The different ecologies of homes made standardization of house dust extract difficult, even though ads like this one for a universal allergenic extract of purified house dust from Endo Products claimed otherwise. Courtesy of Endo Pharmaceuticals.

Figure 37. Dusting was an occupational hazard of the housewife and reflected the gendered spaces of the postwar American home. Reproduced with permission by Pfizer Inc. All rights reserved.

not to the lab scientist but to the housewife. In the gendered space of the home, women, noted the Chicago physician Morris Fishbein, were far more "likely to be in contact with dust than [were] men."[37] Since the average metropolitan home collected over forty pounds of dust per year, dust represented a significant occupational hazard, not only to the housewife, but to her family as well.[38] Its management rested largely in her hands. Even as "time-saving" household appliances—refrigerators, electric washing machines, and vacuum cleaners—became common fixtures in almost every American home, management of the household in the postwar years remained a labor-intensive, gender-segregated task. In 1965, the average American housewife spent over seven and one-half hours per day cooking, cleaning, laundering, and caring for children. These figures differed little from similar time-management surveys of the home made thirty years earlier.[39] Women, in short, spent a lot of time inside the home. They knew this indoor environment far more intimately than men. And doctors and manufacturers targeted the expertise and purchasing power of women in making the home safe from allergy (Figure 37).

As house dust became a visible and known allergen, management of the allergic home was based on the dust-free interior design of antiseptic America that had evolved in response to the germ theory of disease. Since the early 1900s, women had heard, read about, and put into practice what Ellen Richards, the founder of the home economics movement, called "the science of the controllable environment."[40] The message of such sanitary crusaders was clear. The sanctified space of the outdated Victorian home was no safeguard against the dangers of modern life. Germs, harbingers of disease, existed everywhere. They entered through faulty sewage systems; contaminated water supplies; invading insects; and dust that settled on bric-a-brac, carpets, draperies, and furniture. Scientific housekeeping, informed by the principles of modern bacteriology, offered the surest protection of one's family from the dangers of disease.[41] Cleaning was a noble scientific calling, "a fine action," Ellen Richards exclaimed, even "a sort of religion, a step in the conquering of evil."[42] No greater evil existed than the cluttered, wall-papered parlors of the Victorian era, against which home economists and domestic advice experts railed in the pages of *Ladies' Home Journal, Good Housekeeping,* and other popular women's magazines. Sanitary tiles and smooth wooden floors, clean straight lines, painted walls, and a minimalist aesthetic—this was the new style of the modern, healthful home. On polished chrome and white tile, on leather furniture and stainless steel countertops, dust, that "friend of disease," could not so easily hide.[43]

By the 1930s, advice columns and articles in mass-circulation periodicals such as *Better Homes and Gardens, Hygeia,* and *Scribner's Magazine* recommended environmental control of the home as a principal line of attack for sufferers plagued by allergy year round. In 1938, *House and Garden* featured a hypoallergenic Brooklyn apartment. The challenge was considerable: to "make the apartment impervious to outside dust"; to eliminate any dust that might arise inside from rugs, fabrics, and other furnishings; and to create "an inviting, homelike interior for the occupants." Air conditioners and filters were placed in each window. Rugs, draperies, and curtains were out. Venetian blinds and architectural valences to decorate windows were in. Inlaid linoleum replaced carpeted floors. Upholstered furniture was covered in nonallergenic casings, and wooden furniture frames were prepared using a washable finish. The interior designer took care to select "non-pollen bearing plants" to complete the decorations.[44] *House and Garden* did not say what this home makeover cost, but the air conditioners alone put the apartment out of the reach of most middle-class Americans in the 1930s. Nevertheless, more modest interventions could be made to alleviate the sneezing of allergy sufferers irritated by their home environments. A host of products designed to eliminate indoor allergens came on the market, stimulated by a growing "nationwide consciousness of allergy."[45]

One such product was the vacuum cleaner, the first modern weapon in the arsenal of domestic science's war on dust. Doctors used it as a scientific instrument to sample the home and prepare allergenic extracts to use in desensitization treatments. Women were instructed in the art and science of its use by home economists, physicians, and door-to-door salesmen. Unlike dry mops and brooms, which stirred up clouds of dust and exposed the housewife and her family to countless unknown irritants and microbes, the vacuum cleaner sucked up and trapped the hazards. In fact, the modern-day electric upright version of this household appliance can be traced to the keen detective powers of an asthmatic.

In 1907, James Murray Spangler, a seventy-one-year-old janitor in Canton, Ohio, began to notice that his asthma always got worse when he swept the floor of the dry goods store where he worked. Something in the dust he breathed aggravated his cough and wheezing. A handyman by trade, Spangler set to work to design a sweeper that would remove the dust rather than stir it up in the air. He cobbled together a tin soap box with an electric motor fan and pillow case as a dust collector and attached a broom handle to it. Not only did his "electric suction sweeper" work, but it also had, he realized, great sales potential. Spangler took out a patent and in 1908 sold the invention to William H. Hoover. Hoover, the owner of a saddlery company, recognized the sweeper's profit potential when his wife, a friend of Spangler, brought the machine home, having been impressed by its cleaning powers. In 1908, the Electric Suction Sweeper Company (which changed its name to the Hoover Company in 1922) marketed the first electric sweeper in the United States.[46] In 1929, American consumers purchased a staggering 1.25 million vacuum cleaners; the electric sweeper was fast becoming the staple of every allergic and nonallergic household. By the late 1930s, companies like Rexair were offering even more specialized vacuums to allergy sufferers. They used water rather than a cloth bag to trap dust (Figure 38). In 1954, when vacuum cleaners were in use in well over 50 percent of electrified homes in the United States, the Hoover Company was a $52 million business. Door-to-door salesmen touted the virtues of their particular brands, but all relied upon a pitch that sold the product as a necessity in the maintenance of a modern, healthy home. Sales of the Kirby Home Sanitation System, for example, included a demonstration in which the salesman would vacuum the mattress of a prospective buyer's bed. The dirt on the test cloth, the salesman explained, came from "body ash"—the skin shed during the course of a night's rest—and from body secretions and "invisible germs." For added effect, the sales pitch mentioned that the germs would work their way into the mattress and "breed and grow to reinfect unsuspecting people."[47] As the place where an individual spent one-third of each day buried in a pillow, inhaling the dust from feathers, shed skin, cosmetic products, and other suspected allergens, the bedroom became a particular focus of attention in controlling the home environment. The vacuum cleaner offered the first line of defense.

By World War II, the smart consumer could choose from a range of products in securing the home against the allergic patient's worst indoor enemy: dust. "Scientific bedding products," from "Allergi-Rest" pillows to allergen-proof rubberized mattress and upholstery covers, could be purchased in major department stores or through mail-order catalogs. Cosmetic companies began producing makeup lines that provided "non-

Figure 38. Specialized vacuum cleaners, like Rexair, in addition to other home care and cosmetic products, began to appear in the 1930s to help allergy sufferers build a breathing space within the home. Reprinted with permission of Rexair LLC.

allergenic protection and feminine standards of appearance" for women sensitive to orris root, a highly allergenic natural ingredient derived from iris plants and found in practically all cosmetic powders, lipsticks, creams, and fancy soaps.[48] Advances in synthetic chemicals, the results of wartime research, also offered new hope in allergy-proofing the home. In 1946, Karl Figley speculated that the demand for foam rubber for nonallergenic mattresses and upholstered furniture would remedy diminished production in the country's synthetic rubber plants. With doctors recommending that parents send their asthmatic children off to camp with sponge rubber pillows, the market seemed bright.[49]

No allergy product better epitomized the postwar sentiment that technology could engineer a better home and a better world than did Dust-Seal. World War II gave birth to myriad man-made substances that were introduced into the environment allegedly to improve Americans' quality of life. Dust-Seal was one. In the fall of 1947, Arthur Coca, founder of the *Journal of Immunology* and medical director at Lederle

Laboratories, contacted Leonard S. Green, head of Air Sanitation Products. He wanted to talk about a product first developed by the military during World War II to reduce respiratory infections in soldiers by eliminating bacteria-carrying dust in army barracks, military hospitals, and other confined quarters. An adsorptive oil that could be sprayed on blankets, mattress covers, sheets, pillows, and floors enabled the military to successfully reduce bacterial counts between 50 and 90 percent at Camp Carson, Colorado. Green's company had just created a version of this "microbial glue" for civilian use. Coca was greatly interested in experimenting with it in the treatment of allergy.[50]

Although Coca was head scientist at a leading pharmaceutical firm, his interests in the product were largely personal. Coca was plagued by severe migraines, dizziness, and hypertension in his later years and became convinced that his symptoms were the result of allergies to food, dust, tobacco smoke, and other substances, even though he did not display the usual wheal-type reaction in standard scratch tests. Coca found other patients with similar symptoms and puzzling sensitivities, which he argued could be detected through elevations in their pulse. This phenomenon, which Coca labeled idioblapsis and which bears striking similarities to more recent, equally contested, theories of multiple chemical sensitivity syndrome, led him to argue for a much more expansive definition of allergy than the established medical profession was willing to recognize. Although a luminary in the field—he coined the modern term "hypersensitivity" and co-authored the earliest and most authoritative textbook on asthma and hay fever—Coca found himself on the margins of the profession he had helped establish on account of his new-found illness and theory.[51]

Coca went to extreme measures to isolate himself from suspected allergens in his environment. He built a special semidetached kitchen in his house so that he would be removed from the smell of cooking gas. In it, his wife, using only nonaluminum cookware, prepared him a special diet free of milk, peanuts, wheat, and other foods. In the oven, she also baked her husband's newspapers and magazines to protect him from reactions to newsprint and paper lint. He slept in a partially closed-off sunporch. After all this, Coca noticed that every time the house was vacuumed, he was overcome with dry mouth, diarrhea, and a high pulse rate. This realization led him to install a special air filter and add dustproof covers to his bedding and to most of the upholstered furniture to make his home dust-free. While these measures enabled him to move back into the house, Coca still found it necessary to leave the home on cleaning day.[52]

Green's product seemed to be the solution for which Coca was searching: a way to completely seal out dust from his home. After spraying all the rugs with Dust-Seal, Coca reported that he could stay in the house even in the midst of vacuuming and with the air filter turned off. He quickly tried the product out on other patients sensitive to dust. Some of these patients had displayed classic allergic symptoms, such as asthma and chronic rhinitis, whereas others had shown more nebulous idioblaptic symptoms such as headaches, nervousness, and gastrointestinal disorders. When a Teddy bear that always triggered an asthma attack in a four-year-old boy was doused with Dust-Seal, the boy found he could snuggle with his favorite stuffed animal and remain wheeze-free. Other reported cases of success led to both ringing endorsements by physicians and widespread use. *Newsweek* and *Better Homes and Gardens* heralded Dust-Seal as a "definite step in the right direction" in making the home safe for an estimated 5–15 million victims of house-dust allergies. At $4.25 a pound, which was enough to treat just about every possible source of house dust in one room, Dust-Seal appeared to be an affordable and simple technological solution in the fight against dust (Figure 39).[53]

In the 1950s, the modern housewife, equipped with the latest technology and proper scientific understanding, could confidently build a safe place for her allergic loved ones in the confines of the home.[54] It might take a great deal of care and sacrifice. As the baby-boom decade opened, Dr. Jerome Glaser warned that the parents of an asthmatic child "must be prepared to accept the fact that, as in any other chronic disease, the whole design of family life must revolve around the afflicted child. This may not be easy. The mother whose heart is set on loading her home with overstuffed furniture, draperies, and other furnishings likely to produce or hold dust, and the father, whose hunting dogs are his pride and joy, must adjust themselves to the sacrifices made necessary by the child's illness."[55] Such adjustments involved far more than getting rid of the favorite household pet, stuffed animal, or upholstered chair. More than walls made a home. Emotions too, Glaser reminded parents, were triggers equally important as dust in the home environment, and they could spark an asthma attack. In advice columns, magazine articles,

Figure 39. Like Dust-Seal, Allergex offered the modern homemaker a simple and affordable technological solution in the fight against dust. Courtesy Hollister-Stier Laboratories.

and doctors' offices, mothers learned that too much love or too little could exacerbate the symptoms of their asthmatic child. In the heyday of psychosomatic theories of asthma, mothers of allergic children found themselves being asked to take charge, not only of the home's physical environment, but of its emotional one as well.[56] "Any upset in the atmosphere of the home" needed to be avoided, and expressions of anxiety, as well as excessive sympathy around the afflicted child, needed to be controlled.[57] Physically and emotionally secure, in the comforts of home, the white, middle-class asthmatic child of America's baby-boom generation grew up in a refuge, sheltered from indoor dust and from the anxieties and dangers of the outside world.

"Home sweet home it may be," *New York Times* journalist Ralph Blumenthal reported in the fall of 1980, "but the coziest nest can mask a household of hazards, recent Government reports suggest." Blumenthal quoted Susan B. King, chair of the Federal Consumer Products Safety Commission, who sounded an early warning to the American public about the numerous risks they faced inside the home. "You're not talking about a loose roller skate on the stairs," she said. "You're talking about sophisticated problems, unknown hazards—overloads from high-tech appliances, toxic fumes from burning plastics and non-natural fabrics and insulation we all thought was good."[58]

Something had changed dramatically in little more than a decade in the way Americans perceived their homes. The engineered indoor environment, previously touted as a solution to the menacing threat of outdoor air pollution, had suddenly become a public health concern. Why the sudden shift? It was an unexpected outcome of technological solutions to yet another environmental problem: the oil crisis of the 1970s.

The indoor life—at least the particular form it took in postwar America, with all-electric suburban tract homes connected by grids of power lines and highways to urban power plants, shopping malls, office parks, and city centers—was a costly affair.[59] America's thirst for oil more than doubled between the close of World War II and October 1973. That month, Arab members of the Organization of Petroleum Exporting Countries (OPEC) announced an embargo on petroleum shipments to the United States and other Western nations that had supported Israel in its conflict with Syria and Egypt. As pump prices and home heating costs skyrocketed, Americans looked to tighten their pocketbooks by sealing their homes against the costly loss of heat to the great outdoors. Home energy conservation was more than a cost-saving measure. With over one-third of U.S. energy consumption spent on heating and cooling, conservation formed an important national strategy under President Jimmy Carter to lessen American dependence on foreign oil. By the late 1970s, the majority of states had adopted building codes that conformed to revised standards proposed by the American Society for Heating, Refrigeration, and Air-Conditioning Engineers. These new standards adopted minimal ventilation rates as the new recommended standard. Sales of urea formaldehyde home insulation, sprayed into the leaky walls of residences, soared. So did sales of other insulation and energy-saving products, like caulking and weather-stripping. Such energy saving measures reduced outdoor air infiltration and exchange by four to ten times that found in an average leaky home. Demand for alternative fuels also greatly increased the sales of wood-burning stoves and kerosene heaters. In 1981, Americans purchased 1.5 million wood-burning stoves, over seven times the number sold before the OPEC embargo.[60]

Tightly sealed homes now tightly contained the vapors given off by synthetic building materials like particleboard and wall-to-wall carpeting, as well as any household chemicals, tobacco smoke, and combustion products of gas cook stoves and heaters that might be present. The home environment, once considered a safe haven, became an ecological nightmare. Americans spent over 90 percent of their time indoors and the majority of that time at home. Concentrations of some contaminants, including nitrogen dioxide, carbon monoxide, formaldehyde, radon, and asbestos, far exceeded national ambient air-quality standards. The Committee on Indoor Pollutants of the National Research Council, the research arm of the National Academy of Sciences, concluded in 1981 that indoor air pollution had become a "national health concern."[61] The report came out at the very moment that Ann Gorsuch, head of the EPA, slashed her agency's budget for research on indoor air pollution. Connecticut senator Toby Moffett, chairman of the House Energy and Natural Resources Subcommittee, scolded Gorsuch for failing to deal with the health hazards of indoor pollution. But the potential liabilities at stake for corporate America—including the building industry, carpeting manufacturers, and big tobacco—were huge if indeed their products were releasing contaminants that were found to have adverse health effects. Moffett, who got his taste for consumer activism working for Nader's Raiders and organizing the Connecticut Citizens Action Group in the 1970s, was well aware that it was a Pandora's box he was pushing to open. Gorsuch was determined to keep it closed.[62]

As the issue of indoor air pollution heated up in the early 1980s, allergies became a hotly contested political and scientific issue. The 1981 report by the National Research Council's Committee on Indoor Pollutants was not the first to raise concern about indoor air quality. Almost twenty years earlier, Dr. Theron Randolph, a Chicago physician specializing in allergy, had published *Human Ecology and Susceptibility to the Chemical Environment,* which questioned the extent to which humans as biological creatures were capable of adapting to the "ever increasing role of synthetic chemicals" in their lives. Randolph speculated that patients with multiple subjective symptoms—chronic fatigue, migraines, gastrointestinal disorders, asthma, and depression, among others—were actually victims of a susceptibility to countless new chemicals that were being added to the environment on a daily basis after World War II.[63]

Take, for example, Mrs. B. A former cosmetics saleswoman, Mrs. B first visited Randolph in 1947, complaining of hay fever, asthma, frequent headaches, hives, fatigue, and chronic depression. When her illness grew progressively worse, despite the elimination of house dust, silk, and certain foods to which she was allergic, Randolph began to suspect that he was seeing the tip of an iceberg. Hidden and largely undetectable factors in the environment—indoor chemical contaminants, chemical additives in food, synthetic drugs and textiles—he believed, were unsuspected sources of exposure that contributed to chronic debilitating illness. The effects of these factors could easily be confused with or masked by known allergens. The exacerbation of Mrs. B's hay fever and asthma during the summer months, for example, was not unusual—except for one fact: her symptoms did not correspond to the local hay fever season. Randolph traced her sneezing, coughing, and wheezing, not to the outdoor air of the summer home to which she migrated at the same time each year, but to the home's indoor pine paneling. A favorite upholstered chair and dust magnet in Mrs. B's home had also been a source of consternation. So she had sprinkled it with a liberal dose of Dust-Seal, only to find that the chemical product provoked an asthma attack more readily than did the dusty chair. Puzzled by this turn of events, Randolph discovered that it was not house dust but rather the phenol preservative in the house-dust extract used for allergy testing to which Mrs. B reacted.[64]

Randolph was not alone in wondering about the consequences of living in an increasingly "chemically contaminated world."[65] Warren Vaughan, a leading allergist in the United States, speculated on the matter in his 1941 book, *Strange Malady,* which was promoted by the American Association for the Advancement of Science. He wondered if the proliferation of chemicals being introduced into the home, workplace, and larger environment, combined with the hectic pace of modern society, had led to an increased hypersensitivity to a multitude of "strange new substances" that were previously unknown to humans and different from the biological allergens with which most physicians were familiar.[66] But shortly after Randolph established the Society of Clinical Ecology in 1965, a remarkable discovery in the field of immunology pushed his theory of chemical hypersensitivity to the margins of medicine. In

1967, Drs. Kimishige and Teruko Ishizaka, a husband-and-wife research team working at CARIH in Denver, identified immunoglobulin E (IgE) as the key antibody in the allergic response.[67] Now allergists had a clearer basis for understanding the allergic reaction at the molecular and cellular levels. They also had a biochemical measure to identify and more precisely define allergic disease. Patients displaying multiple chemical sensitivities had no detectable IgE in their bloodstream, indicating the lack of a mediated response to one or more offending agents. After this discovery medical specialty organizations such as the American Academy of Allergy, the nation's largest group of allergists and immunologists, dismissed Randolph's views and his patients' symptoms as having no biological basis. Mainstream allergists typically did not deny the complaints of chemically hypersensitive patients. They merely attributed them to the mind rather than the body and sent patients with multiple symptoms off to consult a psychiatrist. Meanwhile, a small but growing number of clinical ecologists, some of whom were board-certified allergists and members of the American Academy of Allergy, embraced Randolph's theories and, outside the limelight of the medical profession, prescribed treatment by means of special diets and chemically free environments.

Once indoor air pollution became a topic of media interest and public concern, however, clinical ecologists fought hard to gain public and legal recognition for their treatments and theories. In the eyes of clinical ecologists, the National Research Council's 1981 report on indoor air pollution gave scientific legitimacy to what their patients' bodies had been saying all along. If, as the nation's leading scientists acknowledged, very small concentrations of formaldehyde in mobile and conventional homes could produce dizziness, headaches, asthmatic symptoms, and eye and throat irritations, then what about all the other chemicals pouring into the environment? One estimate, made by MIT chemist Nicholas Ashford, put the size of this chemical stream at 108 million tons in 1985. "The human body just wasn't biologically designed to contend with the synthetic chemicals that surround us," Randolph told *Glamour* magazine in 1981. "After too much exposure, its ability to handle them simply breaks down."[68] On television and radio talk shows and in newspapers and popular magazines, the public learned of a new disease that went by various names: chemical allergy, environmental illness, total allergy syndrome, chemical hypersensitivity syndrome, and twentieth-century illness. It also learned about a new brand of doctor—the clinical ecologist—who could help those who reacted to "the very fabric of modern life."[69] Stories of people confined to a life of isolation in remote desert locations or in apartments stripped bare of modern conveniences—gas stoves, perfumes, carpeting, packaged food, anything with traces of synthetic chemicals—frequently appeared in the popular press. In 1981, the California legislature took a step toward legitimizing multiple chemical sensitivity (MCS) when it introduced a bill that would establish an advisory committee to serve as an educational clearinghouse and to facilitate research on the environmentally ill. Although it passed in both houses of the California assembly, the bill was defeated by the stroke of Governor George Deukmejian's veto pen.

MCS victims and their doctors pushed their cases in the courts. They hoped to win legal recognition of the disease, which would entitle sufferers to receive Medicaid or Medicare coverage, worker's compensation, and disability benefits. Such recognition would also enable them to seek compensation from chemical companies. But the AAAI, as well as the chemical industry, came down hard. Using their power and influence, these potent groups launched a vehement attack on clinical ecologists.[70] The Chemical Manufacturers Association (CMA), a trade organization in Washington, D.C., representing 182 companies in the United States, warned its members of the "potentially enormous cost that could accrue" if judges granted legal standing to environmental illness. The CMA acknowledged that people diagnosed with "environmental illness clearly merit the compassion and understanding of the medical and social communities." But it emphasized that such medical attention be "placed on proper psychological diagnosis and treatment rather than upon false labels and therapy." The best media strategy, the CMA advised the chemical industry, was to leave the debate in the hands of physicians, who were on their side. In cases where environmental illness surfaced as a legislative issue, the CMA recommended "coalition with the state medical association."[71]

Of the medical organizations that came to the defense of the chemical industry, the AAAI stood out. Under Raymond Slavin's presidency, the AAAI fought a vigorous public campaign to reject the claims of clinical ecologists. In a call to arms sent to its 3,800 members on 1 November 1983, Slavin laid out a plan for the fight against clinical ecology. One action point included a sample letter that doctors were to encourage their patients to send to the Health Care Financing Administration (HCFA), which was considering making diagnostic and treatment techniques commonly used by clinical ecologists ineligible for Medicare reimbursement. The sample letter began, "I have (my child has) gone to doctors who use the technique that HCFA is planning to exclude from Medicare. They were untrained and exploited me (us)." The gloves were off. A lawyer for the Society for Clinical Ecology, Richard Spohn, sent Slavin a letter suggesting that the AAAI's campaign was "motivated not by a fidelity to science or healing, but rather to economic self-interest."[72]

Even Ann Landers got caught up in the dispute. In August 1986, the AAAI published a position paper stating that no factual basis for environmental illness existed and "no immunologic data to support the dogma of clinical ecologists" had ever been produced.[73] Two months later Landers published a letter from "Daisy in Dayton," an alleged sufferer of this supposedly nonexistent malady. Daisy reported that for years she had suffered a "living hell because of depression." A physical exam revealed Daisy was allergic to tuna fish. When she eliminated it from her diet, Daisy reported her life "changed completely." "I look forward to each day. I am off all medication and feel like a new person," Daisy exclaimed. Ann advised her readers that Daisy's experience was not uncommon. Food allergies could readily be a cause of depression, she stated. Ann went on to urge readers who suspected food might be the cause of their psychological problems to consult a clinical ecologist. It didn't take long before Ann's desk was inundated with letters from members of the AAAI. Bruce Berlow, an allergist at the Sansum Medical Clinic in Santa Barbara, castigated the domestic advice expert for "advocating a fringe cult medical group to the public." "You have joined the numerous others whose M.D. degree was obtained at B. Dalton's," Berlow quipped. "I am ashamed at your performance," he concluded.[74]

The dispute over environmental illness revealed how differently clinical ecologists and allergists thought about exposure inside the home. Allergists didn't deny that a relationship existed between indoor pollutants and health. Something in the air accounted for the increasing burden of asthma on society. In economic terms, the costs in 1992 amounted to $6.2 billion—a 39 percent increase over similar estimates in 1985. From 1980 to 1994, the self-reported prevalence rates of asthma had jumped 75 percent. Mortality rates for asthma were also on the rise, especially among African Americans. Genetics couldn't explain a rise so rapid. Even more puzzling, the escalating trend occurred despite better drugs and better outdoor air quality in certain areas. Two factors, however, coincided with the growing mortality and morbidity of asthma: tighter homes and increased time spent indoors. The home environment came under intense scientific scrutiny as researchers sought to understand the role of exposure to indoor allergens and chemicals in what appeared to be an emerging epidemic of allergy-related illness. The key question was what indoor pollutants mattered most.[75]

Ever since allergists had first partnered with engineers to build a breathing space inside the home, the focus of attention had been on keeping nature out. Air filters and air conditioners cleansed the outside air of pollen, the first major allergen that doctors had identified as a contributor to hay fever and asthma. When allergists wanted to detect exposure, they looked first and foremost to the biological world: pollens, arthropods, insects, animal dander, fungi, and other creations of nature. Proteins, not chemical poisons, were the usual suspects in their search for allergens. When the Institute of Medicine, the medical advisory arm of the National Academy of Sciences, outlined a major research study in 1988 to address the problem of human exposure to indoor allergens, biological agents were at the top of its list. The study largely excluded chemical allergens. Most allergists considered chemicals to be a problem of occupational, not residential, exposure. Also, well aware of the debate with clinical ecologists, the National Academy of Sciences did not wish to give any attention to the issue of MCS, a condition for which it found no convincing scientific evidence of an immune system reaction.[76] Allergists were not much interested in the chemical changes to the indoor environment wrought by the building industry in response to the 1970s energy crisis. Rather, they were most interested in the ideal ecological conditions that had been created for a certain home companion: the dust mite.

The dust mite can be an intimidating creature, especially when magnified three hundred times (Figure 40). Blown up to such proportions, this member of the spider family looks more like a creature from a 1950s science fiction film than a common resident of the home. Today the dust mite is widely recognized as the biggest cause of allergic reactions in the home. But prior to the 1980s, few Americans knew that such a creature slept with them at night, defecated in their carpets, and bred in their upholstered furniture. While scientists had known since the 1960s that

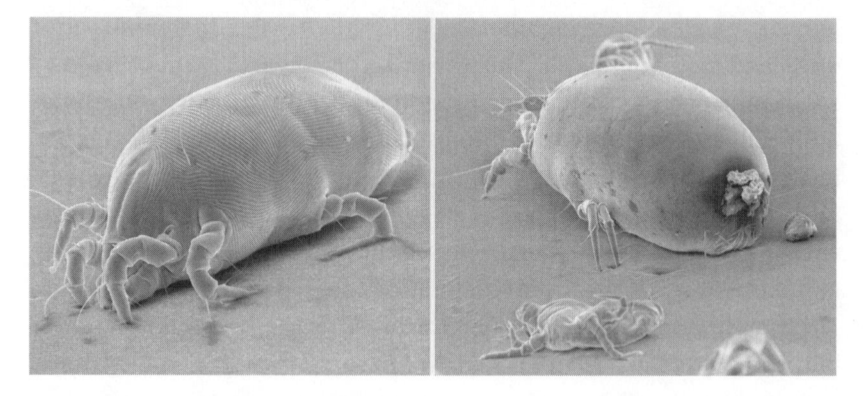

Figure 40. The dust mite, *Dermatophagoides farinae.* Copyright Acarology Laboratory, Ohio State University.

the dust mite was a potent indoor allergen, it was during the 1980s that the ecology of the indoor environment and its effects on health threw the dust mite into the public limelight. Through the ecology of the home, the dust mite became visible and an object of scientific study.

When allergists first identified house dust as an allergic trigger, they also learned that its allergenic properties were highly variable. House dust could be potent or weak in its effects on allergy sufferers, depending on where the dust samples came from. In the 1920s, the Dutch researcher Willem Storm von Leeuwen and his colleagues had noticed differences in dust. They found that dust collected from houses built in areas with heavy clay soils and high underground water tables was far more reactive than dust from homes built on sandy soils or in higher elevations. The dust's potency, these Dutch researchers observed, also fluctuated with changing seasons. Whatever allergen existed in house dust, it clearly depended on temperature and dampness. To a later generation of Dutch researchers working in the Department of Allergology at the University Hospital in Leiden, such fluctuations suggested a biological cause. In the early 1960s, Reindert Voorhorst, a specialist in internal medicine, teamed up with Hendrik Varekamp, a former colleague of Storm von Leeuwen's. They hoped to determine once and for all the mysterious source of the house-dust allergen. The researchers began by classifying the different parts of the city of Leiden by the quality of housing stock. As a result, they found that the rate of asthma among younger patients who were sensitive to house dust and who grew up in poorly built older homes was almost double that of patients in the same age group who were raised in newer homes. In comparing the different qualities of housing construction, the team found that the older homes had much higher levels of dry rot. Once again, dampness appeared to be a critical factor in the amount and potency of house-dust allergen inside different homes. Voorhorst added an ecologist to the team of house-dust scientists. Marise Spieksma-Bozeman, a Master's student in botany and zoology at the University of Leiden, quickly found in the house-dust samples something that had evaded researchers for forty years. She observed a previously unknown mite living among the scales of human skin, animal dander, fungi, and pollen in a miniature living ecosystem. She also noted much higher populations of the mite in the dust samples collected from damp homes. In 1966, this new species of mite, *Dermatophagoides pteronyssinus,* was named and its allergenic properties identified. The mystery of house dust's allergenic properties had finally been solved.[77]

Over the next decade, scientists learned more and more about the intimate life and ecology of the dust mite. Two of the most cosmopolitan species, *D. pteronyssinus* and *D. farinae,* found in carpeting, household furniture, and beds, as well as in the nests of birds and mammals and on human and animal skin, live in Europe and North America. In the United States, *D. farinae* is the more common species. Its favorite food is the dead scales of human skin; the average person sheds a gram of skin per day, a more than ample supply to sustain a thriving population of mites. But food isn't the limiting factor in the ecology of the dust mite. Humidity is. The dust mite's optimal comfort zone is 60 percent humidity at 70 degrees Fahrenheit. If it gets much drier than that, the dust mite has difficulty surviving for two reasons. First, the mite cannot cope with moisture loss. Second, the lower levels of relative humidity are also detrimental to certain species of fungi that the mite uses both as a food source and as a help in digesting human skin. Because of its particular needs, the dust mite finds the bed to be a very comfy place: it is a warm, moist environment with lots of food nearby. Carpets and upholstered furniture are other favored niches. Under the right conditions, one gram of house dust can be home to one thousand mites. Concentrations of more than five hundred mites per gram of dust can be a serious risk factor for asthmatics. But the dust mites themselves are not the source of the problem. It's their feces. Under optimal environmental conditions (for mites), a gram of dust can contain over five hundred thousand fecal particles. Of the allergic patients who receive desensitization treatments in the early twenty-first century, more get injections of house-dust extract purified from dust mites and their excrement than any other type of extract.[78]

Why did the dust mite burst onto the scene only in the mounting concern over indoor allergens in the 1980s when allergists had known of its presence for some time? Again, the explanation lies in energy conservation strategies in the building industry. Tighter buildings not only locked in indoor contaminants but also locked in the moisture in which dust mites thrive. In Denmark, for example, where building codes take into consideration a regulation of the indoor climate for allergy sufferers, researchers found that the tightening of apartments through standard energy conservation measures reduced air infiltration and energy consumption, but it also increased indoor humidity and dust-mite populations. In the United States, the Committee on the Assessment of Asthma and Indoor Air of the Institute of Medicine reported in 2000 that in homes or schools with evidence of dampness problems, prevalence rates of asthma increased by a factor of two. The leading suspected cause was the dust mite. Indeed, the committee report stated that of all indoor biological allergens listed (including cats, dogs, cockroaches, and molds), only dust mites were considered to have a known causal relationship with the development of asthma. The convergence of more tightly sealed homes, more time spent indoors, and the ecology of the dust mite led allergists and the popular press in the 1980s to seize on dust mites as an explanation for the steady rise in childhood asthma.[79]

Invisible to the naked eye, the dust mite initially attracted far more attention in research on indoor allergens than another common but visible fixture of the home: secondhand tobacco smoke. Long before Surgeon General C. Everett Koop alerted the public to the health dangers of environmental tobacco smoke in 1986, clinicians were well aware of allergy sufferers' complaints about the irritating effects of tobacco smoke. Katherine Bowman-Walzer, a research assistant in Arthur Coca's lab, sadly recalled how the close personal and professional relationship she had with her mentor grew more distant in the 1940s because of her own smoking habits. As Coca's health deteriorated and his preoccupation with chemical hypersensitivity grew, he increasingly complained of the smell of tobacco smoke that clung to Bowman-Walzer and allegedly exacerbated his allergic symptoms.[80] But the solid particulates and chemicals that permeated the air with every puff of a cigarette were chemical exposures that had little history in scientific research on allergic disease. What allergists were willing to recognize as visible allergens inside the home was largely determined by biochemical assays at the cellular level and the presence of IgE and other products of the allergic response. The isolation and purification of dust-mite allergen and the biochemical pathways it initiated inside the body could be seen easily. But a single exhalation following a drag on a cigarette released more than thirty volatile organic compounds. The causal links, if any, between secondhand tobacco smoke and asthma were clouded in a fog of uncertainty. Furthermore, for allergists to emphasize the role of secondhand tobacco smoke—and thereby chemicals—in an immune-related illness like asthma would indirectly play into the hands of clinical ecologists. Instead, allergists played into the hands of the tobacco industry.

Pressure built in the late 1980s to enact legislation to reduce the threat to human health posed by exposure to indoor air pollution. Fearing regulation targeted specifically at secondhand tobacco smoke, the tobacco industry established the nonprofit Center for Indoor Air Research (CIAR). Only the federal government surpassed the center in terms of money spent on indoor air pollution research. The tobacco industry had good reasons to back a research center for indoor air pollution. As more indoor air contaminants were discovered and as more of the complexities of their interactions and their health effects came to light, the harder it would be to single out secondhand tobacco smoke as an indoor pollutant in need of regulation.[81] Although the tobacco industry embraced uncertain findings, allergists and immunologists stayed clear of inconclusive research. Scientific papers that reported inconclusive results did little to advance one's professional career.

No area of allergic research appeared more plagued by problems of uncertainty than chemical exposure. The debate with clinical ecology made that clear. Consequently, the effects of secondhand tobacco smoke on asthmatics received relatively little attention in a study launched in 1989 and published in 1993 by the Institute of Medicine's Committee on the Health Effects of Indoor Allergens.[82] But the tobacco industry fought a losing battle against the regulation of environmental tobacco smoke in government buildings, the workplace, and other public spaces during the 1990s. In 1986, fewer than 3 percent of Americans who worked indoors did so in a smoke-free environment. By 1996, more than two-thirds of Americans spent their workday in a smoke-free building. The dramatic change was effected in large part by lobbying groups like the American Lung Association and Mothers of Asthmatics, which urged members to speak out publicly against secondhand tobacco smoke.[83] When the Institute of Medicine revisited the subject of indoor allergens in 2000, it concluded that sufficient evidence of a causal relationship between exposure to environmental tobacco smoke and the exacerbation of asthma did exist. It also found sufficient evidence of an association between exposure and the development of asthma in children of preschool age. The tide of research on the susceptibility of asthmatics to secondhand tobacco smoke had quickly turned.[84]

Between 1987 and 1999, federal agencies spent a total of $1.1 billion in research on indoor air pollution.[85] If the figure seems high, think again. A Department of Energy (DOE) study in 2000 estimated a potential annual savings to the U.S. economy of $1–4 billion for a reduction in allergies and asthma through changes in building design, construction, operation, and maintenance. But who would bear the costs of eliminating building-related risks for allergy and asthma sufferers? The DOE study recommended that such costs be left in the hands of the individual sufferers. They alone bore the responsibility of controlling their private breathing space inside the home.[86]

Indoor air, unlike outdoor air, is regarded in the United States as a consumer choice, not a public good. Despite repeated attempts, efforts to pass a national indoor air quality act in the United States have consistently failed. As Americans awakened to health threats posed by the indoor environment, responsibility for protection from indoor contaminants fell not to the government but to the individual consumer. Industry was more than willing and able to meet new consumer demand.

In the 1990s, the number of Web sites and catalogs touting a barrage of environmental control products—including dust-mite testing kits, bedding encasements, high-efficiency dust bags, anti-allergen vacuum cleaners, and dust-mite destroying powders—had spread "faster than tree pollen," noted *U.S. News and World Report.*[87] Increased media attention on indoor air quality and rising asthma rates created a market that manufacturers were eager to exploit. In the late 1970s, only two portable air cleaners, Pollenex and Ecologizer, were available. Each promised to clear the air of tobacco smoke, odors, dust, pollen, and other indoor pollutants. In 1985, *Consumer Reports* listed twenty-three brands with catchy names like Bionaire, Freshenaire, and NatureFresh; these ranged in price from $20 to $300. While the devices used a variety of technologies, from simple filters to electrostatic precipitation to negative ion-generation, all promised to deliver "the cleanest, freshest indoor air you've ever experienced." It wasn't long before the Federal Trade Commission was filing complaints against some manufacturers for false advertising. Often the devices that failed to deliver on claims were the smaller, inexpensive air cleaners, which *Consumer Reports* found "almost useless" in reducing tobacco smoke, dust, and pollen under household living conditions.[88] Furthermore, the majority of these home-remedy solutions were quick technological fixes. Many of them were more sugar pill than magic bullet. Few of them addressed the larger structural issues in the ecology of building design: moisture problems, inadequate ventilation, and susceptible building materials (such as carpeting) that made happy homes for dust mites and molds.

Amid this exploding marketplace of home care products targeted at allergy sufferers, Nancy Sander founded Mothers of Asthmatics in 1985. A freelance writer and mother of four, Sander had spent six years of her life rushing in and out of hospital emergency rooms in the fight to give her severely allergic and asthmatic daughter, Brooke, a space to breathe. When an experimental drug treatment program at Georgetown University that included managed care helped bring Brooke's condition under control, the daughter got a new life and the mother a new purpose. With "nothing but a writing and mothering background, a broken typewriter and $8 with express instructions from [her] husband not to dip into the family savings or the monthly budget," Sander began a newsletter. The *MA Report* was a resource and consumer clearinghouse for parents to exchange tips, coping strategies, and information on consumer products and drugs useful in the care and treatment of their allergic and asthmatic children.[89] In less than one year, five thousand members had joined Sander's organization. Mothers of Asthmatics was founded on the belief that "consumers, when made aware of their options in asthma management plans, will choose a plan that allow[s] them to gain control over the current problem, monitor and prevent future problems and handle a crisis."[90] With the backing of the American Academy of Allergy and Immunology and the support of pharmaceutical companies and industry, Mothers of Asthmatics expanded its reach in 1989 through the creation of the Asthma and Allergy Network, a patient resource for asthmatics of all ages. Through its newsletter, toll-free hotline, publications, videos, and Web site, the Allergy and Asthma Network/Mothers of Asthmatics, which has grown in recent years to a $2 million organization, takes a decidedly grassroots, consumer-oriented approach to the management and treatment of allergy and asthma.[91]

Since the early twentieth century, Americans have sought to engineer an ideal indoor environment as a refuge from allergic disease. Manufacturers of air conditioners, portable air filters, special vacuum cleaners, and allergy-proof covers have been eager to sell a breathing space made pure through technology. Allergy sufferers, when the price is within reach, have been eager to buy quick technological fixes that could remedy the allergic house. This vision of remedy is built upon the assumption that consumers, particularly women, have both a responsibility and a choice to make a healthy home. It is also a version of the American dream shaped largely by the values of the white middle class.

Not all American citizens can buy $600 vacuum cleaners, install the latest High Efficiency Particulate Absorbing (HEPA) filtration systems, or purchase a home with wood floors. For those who struggle to afford an apartment, let alone own a home, many of the environmental interventions endorsed by industry, physicians, and Mothers of Asthmatics are simply not within reach. Management of asthma on the home front is not a consumer choice for those at the margins of society struggling to survive.

The promise of a technological solution to America's growing allergic landscape offered consumers an illusion that in the home they could engineer nature, bring it under control, and create an indoor environment better than the best places nature had to offer. In this air-conditioned paradise, Americans could find health and comfort 365 days a year. Engineers hoped to eliminate the ecology of the disease by creating a uniform environment indoors. Industry hoped to profit from an expanding market by selling every allergic consumer a universal remedy. Many Americans have bought this idea wholesale. A 2004 University of Michigan study found that parents with asthmatic children eagerly purchased special air filters, allergen-free vacuum cleaners, mite-proof bedding covers, and other products to reduce environmental exposures in the home and thus, hopefully, their children's allergy symptoms. The majority did so, however, with little attention to or knowledge of whether the products would address the specific allergens to which their children were susceptible.[92] A fancy vacuum cleaner, for example, would do little to help a child whose asthma was triggered by secondhand tobacco smoke. Direct-to-consumer advertising sells products, but it does little to help parents understand the unique environmental circumstances of their children's disease. The homes in which people live and the allergies with which they struggle both have ecologies that a one-size-fits-all technology cannot so easily erase.

CHAPTER

6

An Inhaler in Every Pocket

I have them stashed everywhere. One by the front door, one by the back door, one by my bed, one in the glove compartment, one by the sofa, one in my desk at work, and one in my purse.

—"Irene," 1992

Each morning my grade-school-age son, Keefe, and I share in what is a common ritual for millions of people in the United States. As consumers of allergy drugs, Keefe and I reach every day into the medicine cabinet for our respective remedies. Each drug we take promises, in one way or another, to alter the complex immune system relationships within our bodies so that we might never mind the outside environmental connections within the world in which we live. "One tiny blue pill," claims a 2002 ad for the antihistamine Clarinex, offers twenty-four-hour seasonal allergy relief, "wherever you are, whatever your seasonal allergy" (Figure 41). Allergy drugs promise escape from place.

But the body, where allergy drugs work their wonders, is also a place. In today's medical lexicon, atopy refers to a hereditary predisposition to develop hypersensitivity to what are normally innocuous substances in the environment—pollens, foods, pet dander. "Allergy" is the term for this adverse overreaction to an offending substance, or allergen. A chain of immune system reactions occurs during an allergic response. Part of this response includes the production of antibodies, proteins produced by the body's immune system that recognize and help fight infections and other foreign substances in the body. In the case of allergy, antibodies are produced that specifically recognize an offending allergen. In the late 1960s, the detection by Kimishige and Teruko Ishizaka of heightened levels of immunoglobulin E (IgE) in the bloodstream of hay fever sufferers became a way to distinguish allergic bodies from those without allergies.

Allergic bodies are of great interest to physicians, pharmaceutical companies, and marketing firms eager to prescribe or sell the latest drug remedies. One recent advance in consumer marketing is the concept of tribe. Companies no longer just sell products. They sell particular lifestyles, values, and beliefs that are carefully shaped and packaged to give the individual consumer a sense of belonging. "Which iPod are you?" asks Apple. "Which drug are you?" ask pharmaceutical companies as they work to create tribal relations and markets through "disease, patients, and pills," as Greg Critser notes.[1] In the late nineteenth century, wealthy hay feverites found a shared identity defined by disease, class, and place. In the new millennium, drugs rather than place are the common bonds that unite allergy sufferers—at least sufferers with access to the latest medications—and the identity is marketed as much as found. Each year, pharmaceutical companies herald the newest technological advances with offerings of new and better relief.

Like others in our "tribe," Keefe and I reach for our drugs each day. I pull out a purple disk, ergonomically engineered to fit into the contours of my hand, click back a trigger, put my lips around the mouthpiece, and inhale. A sweet taste lingers in my mouth as a fine powder descends into my lungs. The powder is a combination of fluticasone proprionate (a relatively new inhaled corticosteroid used to reduce chronic airway inflammation associated with asthma) and salmeterol (a beta 2-adrenergic bronchodilator that reduces airway constriction caused by muscle spasms). The product, Advair, is GlaxoSmithKline's biggest moneymaker in the respiratory drug category, a therapeutic area that makes up 27 percent of the company's total pharmaceutical sales. In 2001, when Advair hit the U.S. market, sales exceeded $1 billion. The speed at which it was adopted in the United States—3 million prescriptions were written in the nine months following its release in April—made it "one of the most successful pharmaceutical product launches ever." With Advair, GlaxoSmithKline's Web site assures me, I'll be able "to enjoy more days doing what [I] love, without [my] asthma getting in the way."[2]

Keefe is in the GlaxoSmithKline tribe too. In early spring 2000, when he was three years old and living in Berlin in a damp apartment with

Figure 41. Antihistamines like Clarinex offer allergy sufferers the promise of escape from place. Reproduced with permission of Schering Corporation.

horsehair plaster walls and an old carpet, he developed a series of chronic respiratory infections and a case of pneumonia. His German doctor dispensed a syrup of plant extracts and turned to antibiotics only when nothing else would work. Upon our return to the United States that summer, his physician advised that Keefe begin using Flovent, which has the same inhaled corticosteroid found in Advair, as a precaution against the onset of asthma. Given my own history of asthma, my wife and I reluctantly agreed, even though no conclusive data existed on the long-term effects of inhaled steroid use. Each morning and night, Keefe shakes the aerosol inhaler (connected to a spacer to optimize drug delivery), presses down on the aluminum tube, and inhales a dose of the steroid to coat the airways of his lungs. His body helps sustain GlaxoSmithKline's competitive edge in the global marketplace of respiratory drugs. Together, Advair and Flovent made up 20 percent of GlaxoSmithKline's $32 billion in worldwide pharmaceutical sales in 2005. Keefe and his parents can also feel good knowing that the new Flovent HFA inhaler no longer contains a chlorofluorocarbon (CFC) propellant—a part of GlaxoSmithKline's "corporate commitment to the phaseout of CFCs" and an aspect of its environmentally friendly face.[3]

Inhaled steroids and bronchodilators make up two of the three main pillars of modern medicines used by allergy and asthma sufferers in the twenty-first century. Antihistamines are the third. Another part of Keefe's daily routine is to swallow a tiny white pill. Loratadine is now a generic drug, but as the active ingredient of Claritin, Schering-Plough's blockbuster second-generation antihistamine, loratadine made the company more than $15 billion between 1993, when it first launched Claritin in the United States, and 2002, when its patent on loratadine expired.[4] In Keefe's case, the drug helps control the itchiness associated with dermographism. Dermographism is the most common form of physical urticaria, an allergic skin disorder. Drag your fingernail or a pencil eraser over his skin, and in a few minutes red marks clearly reveal where you pressed. Dermographism literally means "skin writing." Keefe has been known to use his arm as a tablet for games of tic-tac-toe on car trips. Once on a camping trip he used his back as an advertising billboard, selling space to fascinated kids who were eager to see their messages magically appear. Antihistamines help Keefe's skin behave more normally (although being out of the ordinary can have its advantages if it makes you popular among your friends).

Almost the entire history of allergy and asthma treatment and modern drug development is represented by Keefe's and my experiences.[5] My use of one drug, tetracycline, an antibiotic prescribed during my childhood to control pneumonia and chronic respiratory infections, remains visible in the yellowish green stains on my teeth. Now those stains frustrate my dentist, who fails to understand why I don't want to use more chemicals to rectify the "problem" and enhance my smile. The cumulative effects of other drugs are less obvious. What impact might years of oral steroid use during my childhood have had on my small physical stature? Or to what extent did bronchodilators push me to experiment with speed during my teenage years? Questions of efficacy, safety, and synergistic effects arise when I consider my son and his relationship to a culture in which prescribed drugs are no longer about survival but about enhancing one's performance in school, work, recreation, and bed.

Keefe and I visit our allergist once a year. As a medical historian with an anthropologist's eye, I see the doctor's office as a research site as rich as any historical archive. At the clinic, Keefe and I first meet with a nurse and take turns at the spirometer, a machine designed to measure lung function. This is one of the important technologies allergists use to diagnose, classify, and standardize asthmatic bodies. Manufacturers have made it a fun game; kids can choose different scenarios to play. Keefe usually chooses the three little pigs. As he blows into a tube that allows the machine to measure how much air he expels and how quickly, a wolf blows down the brick wall behind which the three little pigs hide. Keefe's goal, the nurse tells him, is to knock the wall down completely with the force of one exhaled breath. Next, we are escorted into a room to wait for our allergist to appear. After catching up on events of the previous year and answering queries from me about what's hot in the allergy field, our doctor focuses on the spirometry results.

It is here that the dance begins. We debate what the measurements mean and how aggressive a strategy to take with available medications. If the forced expiratory volume (FEV) values that the spirometer measures drop below 90 percent, my physician will want to up the dose of meds so as to maximize the days we can do the things we love to do, as the ads for Advair encourage. He wants Keefe to have the competitive edge it takes to perform well playing soccer or diving, two sports Keefe loves. His questions about how Keefe feels when playing soccer or swimming suggest how much the culture of sports enters into our doctor's thinking about what constitutes the "normal" lifestyle of the average American child. Indeed, these conversations reinforce what we read in drug trade journals, popular magazines, newspapers, and medical archives: a culture of performance and productivity has been instrumental in the postwar boom in the use of allergy drugs to transform and enhance the bodies of America's allergic citizens.

Keefe and I have conceptions of normality different from those of our allergist. Before we walk into the office and see the spirometry results, both of us are feeling quite well. Technology tells us one thing about our bodies when they are standardized against a population norm, but our experience tells us another. So my allergist and I usually reach a compromise. The course of action we choose is probably less aggressive than he would like but slightly more than is comfortable for me, given the unknown long-term side effects of some of these drugs. We walk out of the office with a handful of prescriptions, some of which will go unfilled during the year. Noncompliance is a huge loss for pharmaceutical companies. So is switching brand loyalty. That is why GlaxoSmithKline wants Keefe and me to feel a part of its tribe.

The body represents the most recent chapter in the endless search for a breathing space free from allergic disease. In the late nineteenth century, wealthy hay feverites looked to nature as an escape from the harried pace of modern life. In the twenty-first century, more than 50 million allergic Americans now turn not to nature but to modern biotechnology to free them from nature's constraints. After World War II, in an age of synthetic chemicals, wonder drugs, and an expanding economy, when allergy had infiltrated all walks of American life, engineering the body in ways that would overcome the peculiarities of place and environmental change became a consumer and corporate dream.

How antihistamines, one of the first miracle drugs in allergy treatment, became common in the bloodstream of Americans and vital to the circulation and accumulation of capital in the drug industry is a story that begins with World War II. The war had a massive impact on American industry and on American life more generally. The drug industry was no exception.

During the height of World War II, between 1943 and 1945, defense outlays hovered at over 37 percent of the nation's gross domestic product (GDP). After the war, military spending declined dramatically. Still, in the early 1950s, at the height of the Korean War, national defense accounted for close to 13 percent of the country's GDP. Big industries closely tied to the military largely ran America. Under wartime pressures, industrial manufacturers made massive investments in plant and production capacity to deliver new products quickly and directly to the government. The demands of war fostered an economy geared toward production at a time when the government's consumptive capacity seemed limitless.[6] The stumbling block to production was not demand but wartime shortages of raw materials. To overcome problems of access to valuable natural resources, industries became much more intensive in their uses of capital and scientific knowledge. The drug industry is the perfect example. During the war years, this industry became more capital intensive, moving from simply producing extracts of raw materials to chemically synthesizing these substances in the laboratory. This change was driven in large part by wartime disruptions of overseas trade. The takeover by Japan of the Dutch East Indies, for example, where large *Cinchona* tree plantations provided a virtual monopoly on the global quinine trade, intensified efforts among the Allies to produce a synthetic anti-malaria drug that could mimic the therapeutic effectiveness of *Cinchona,* used since the seventeenth century to combat malaria. Furthermore, in pursuit of efficiency, the drug companies developed expertise in modern consumer packaging. This change made it possible to bypass the traditional compounding and packaging of drugs by local pharmacists and also reduced the traditional role of physicians and pharmacists in advertising and promoting drugs.[7]

The postwar economy remained dominated by the federal government. Rationing, price controls, and trade restrictions were maintained in a variety of forms well into the mid-1950s under the administration of President Dwight D. Eisenhower. The federal government's postwar economic strategy aimed to prevent the inflation and idle productive capacity that had been so troublesome for the American economy after World War I and had plunged the country into the Great Depression. The strategy entailed engineering a demand-side boom in which the American public would be persuaded to consume the excess supply created by the wartime transformation of industry. As Leo Nejelski, president of a consulting firm to the drug industry, observed in 1947, "Drug manufacturers are preparing for the biggest marketing push in history. They now have the capacity to produce more medicines and sundries than our people can possibly consume."[8] The 1950s became a golden age of not only wonder drugs but also marketing, a decade in which the brightest psychologists were pressed into service to turn humdrum war surplus into consumer must-haves and to apply their new and sophisticated statistical techniques to the problem of gauging "consumer sentiment." The drug industry was no stranger to modern advertising and marketing techniques; indeed, it had helped usher in the age of professional advertising as a consumer-oriented economy had taken hold in American society prior to World War II. And it would aid in the further expansion of "an economy, culture, and politics built around the promises of mass consumption" in the postwar years.[9]

The strategy for postwar economic growth relied on two key principles: keep American workers producing and keep consumers buying. Antihistamines entered into both the production and consumption sides of this equation (Figure 42). In less than a decade after they hit the U.S. market in 1946, promoted as a remedy for allergy, antihistamines became, next to antibiotics and barbiturates, the third most commonly prescribed class of drugs in America. In less than five years after penicillin became available to American consumers in 1944, sales for antibiotics skyrocketed to $82 million. Antihistamines surpassed even this staggering record of growth. Benadryl and Pyribenzamine, the first American antihistamine drugs, made up the bulk of $2 million in antihistamine sales in 1947.[10] Elliott Bowles, president of the Union Pharmaceutical Company, predicted that by 1950 over-the-counter (OTC) and prescription sales of antihistamines would exceed $100 million. Concerns from the military and industry regarding worker productivity largely propelled that engine of economic growth. But the spark that had initially set that engine in motion was the production of a drug.[11]

Soon after the classification of hay fever and asthma as allergic diseases in the early twentieth century, physicians and medical researchers began a concerted search for the underlying biochemical mechanisms responsible for the allergic reaction. Despite the diversity of symptoms that allergy sufferers displayed—some reacted to food, others to house dust; some were prone to hay fever, others to asthma or dermatitis—the condition ran in families, suggesting that allergy had a strong hereditary

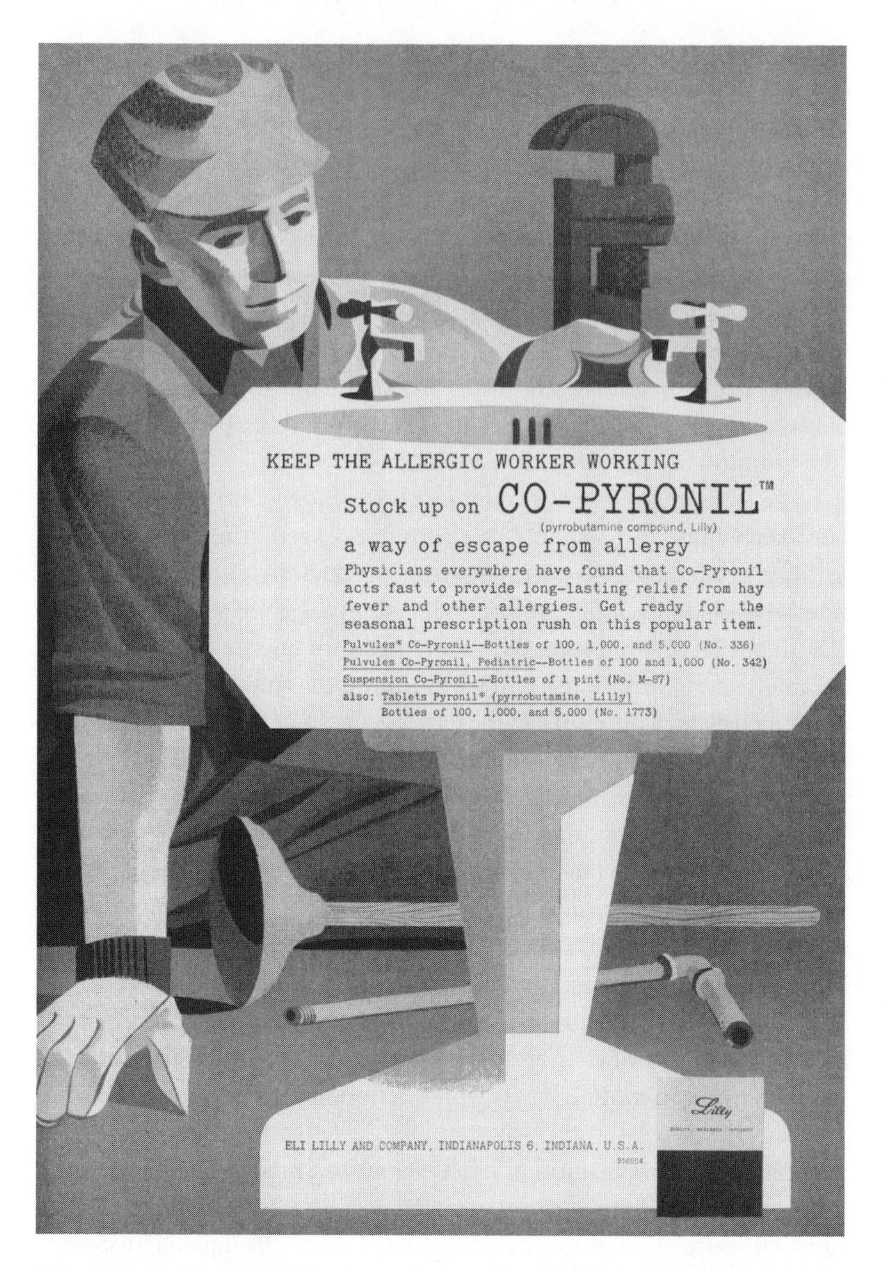

Figure 42. Keeping the "allergic worker working" helped make antihistamines the third most commonly prescribed class of drugs in America during the 1950s.

component. Was there something common in the immunological systems of allergy patients that made them hypersensitive in their own idiosyncratic ways to a range of substances in the environment around them?

In 1921, Carl Prausnitz, a physician and researcher at the University of Breslau, injected into his own abdomen the blood serum from his colleague Heinz Küstner, who had a severe allergy to fish. Prausnitz had no such allergy, but when he sat down to a fish dinner, he found that a case of hives appeared on his body at the injection site. The experiment demonstrated that a specific immune body—later to be known as an antibody—was present in the serum of allergic patients. The substance could be passively transferred to nonallergic individuals, and it played a crucial role in the allergic reaction. What that antibody was and what role it played in the chemical reactions that triggered an allergic reaction remained a mystery. But Prausnitz's discovery suggested that a common biochemical mechanism underlay the widely varying symptoms of allergy sufferers. If researchers could make visible the biochemical pathway of the allergic reaction, they would have both an instrument and a measure to diagnose and standardize allergic bodies, and they might also one day arrive at a universal drug for allergy treatment. Thus began the search for a molecule that might unlock the secret of allergic disease within the body.

Histamine was the first chemical identified as a likely player in allergic reaction. In 1911, while he was the director of the Wellcome Physiological Research Laboratories in London, Sir Henry Dale, who would go on to win the 1936 Nobel Prize in Physiology or Medicine for pioneering work on the chemical transmission of nerve action, demonstrated with his colleague Patrick Laidlaw that the injection of histamine into guinea pigs and dogs caused symptoms resembling those of anaphylactic shock.[12] Over the next three decades, scientific evidence mounted, pointing to histamine, which was found in elevated concentrations in lung, skin, and other tissue of allergic patients, as a potent contributor in the events observed in an allergic reaction. Once in the bloodstream, histamine can wreak havoc. It causes small blood vessels to dilate, which results in the swelling of surrounding tissues and leads to (for example) the nasal congestion experienced by hay fever sufferers or the itchy and blistered skin familiar to people with hives. How histamine worked in conjunction with antibodies and allergens and where it was stored in the body remained a puzzle to researchers at the beginning of the 1940s.[13] But such unanswered questions did not hinder the search for a chemical that could inhibit histamine's action and thereby offer a treatment for allergy. In 1942, the first promising compound, antergan, was synthesized in the laboratories of the Society of Chemical Manufacturing at Rhône-Poulenc in France. Two years later, another French team, working at the Pasteur Institute in Paris, reported favorable results with a related compound, neoantergan, which effectively controlled the symptoms associated with hives, eczema, and hay fever, as well as other allergic conditions, in a majority of patients.[14]

In the United States in 1941, George Rieveschl, a twenty-six-year-old assistant professor of chemistry at the University of Cincinnati, manufactured a new compound in his laboratory that would become an American synonym for allergy relief: Benadryl. Rieveschl wasn't working on creating an antihistamine; he was trying to build a more effective antispasmodic drug. Together with his graduate student Frederick Huber, he manufactured a series of compounds that he sent off to Parke-Davis in Detroit for evaluation. Rieveschl's creation, A524, made its way into a batch of over four hundred compounds that Parke-Davis researchers were testing to counteract anaphylactic shock in guinea pigs. By injecting the chemical just before giving the guinea pig a lethal dose of inhaled histamine, researchers could assess the compound's effectiveness in blocking the action of histamine. A524 proved a winner. Guinea pigs injected with the compound breathed an otherwise lethal dose of histamine and survived, seemingly untroubled by the ordeal. Even before results of the human clinical trials were in, one Parke-Davis chemist downed a couple of tablets of the experimental drug before going to lunch. He hoped to at last enjoy his favorite food—potatoes—without suffering the wheezing and sneezing fit that always accompanied such a meal. Two orders of French fries later, the chemist was still fine. His experience foreshadowed what other experimental subjects confirmed.[15] When Benadryl hit the market in the summer of 1946, just in time for the hay fever season, Parke-Davis reported its "proven effectiveness" in 65–75 percent of hay fever cases. A few months later, Ciba Pharmaceutical Products announced the release of its antihistamine, Pyribenzamine, which it claimed aided 85–95 percent of patients with seasonal hay fever and hives. The antihistamine gold rush was on. By 1950, more than twenty-one antihistamine compounds packaged under one hundred different trade names in tablets, nasal sprays, eye drops, and creams were on the market in the United States.[16]

Initially Parke-Davis and Ciba targeted the seasonal hay fever market in their launch of the new antihistamines. Ads for Benadryl and Pyribenzamine appeared in medical journals and drug trade magazines in the early summer months and continued into the late fall of 1946. Within a year, the two companies were pushing to expand the marketing reach of their antihistamines beyond the hay fever season. "Whatever the source, common allergic conditions—such as urticaria, seasonal allergic rhinitis, asthma—respond favorably to Pyribenzamine hydrochloride," a 1947 Ciba ad proclaimed. To bring the message home, the ad included an image of an arm covered with the wheals caused by allergic reactions to strawberries, feathers, and ragweed. Similarly, Parke-Davis touted in a 1947 ad to drugstore owners that Benadryl was now "prescribed the year 'round," and it listed conditions (ranging from urticaria, eczema, and hay fever to food and drug allergies) for which doctors were prescribing the drug (Figure 43). Penicillin was one of the most widely prescribed of the new antibiotics in the immediate postwar years and was also a common cause of drug allergy; antihistamine prescriptions piggybacked on this rapid rise in prescribed antibiotics.[17]

But the real boom in the antihistamine market came from the military and the problems troops faced combating not the human enemy but the ill winds of nature that brought on hay fever, seasickness, and the common cold. An "ill wind" could blow in "good business," noted a 1947 Ciba ad for Pyribenzamine.[18] And there was nothing better for business than travel sickness and the common cold. Both were classic examples of what is now known in the drug industry as off-labeling: the use of a drug for the treatment of a disorder for which it was not originally developed, marketed, or approved.

In 1947, Leslie Gay, director of the Allergy Clinic at Johns Hopkins University and Hospital, received a number of compounds from G. D. Searle, a family-owned drug company in Chicago, to be included as part of a clinical research study that Gay and his Hopkins colleagues had undertaken on the new antihistamine drugs. G. D. Searle was no stranger to the allergy/asthma market. In 1930, it had begun production of Aminophyllin, a bronchodilator in the same class of drugs as epinephrine, ephedrine, and theophylline. Bronchodilators were the only effective class of drugs available to asthmatics before World War II; they reversed

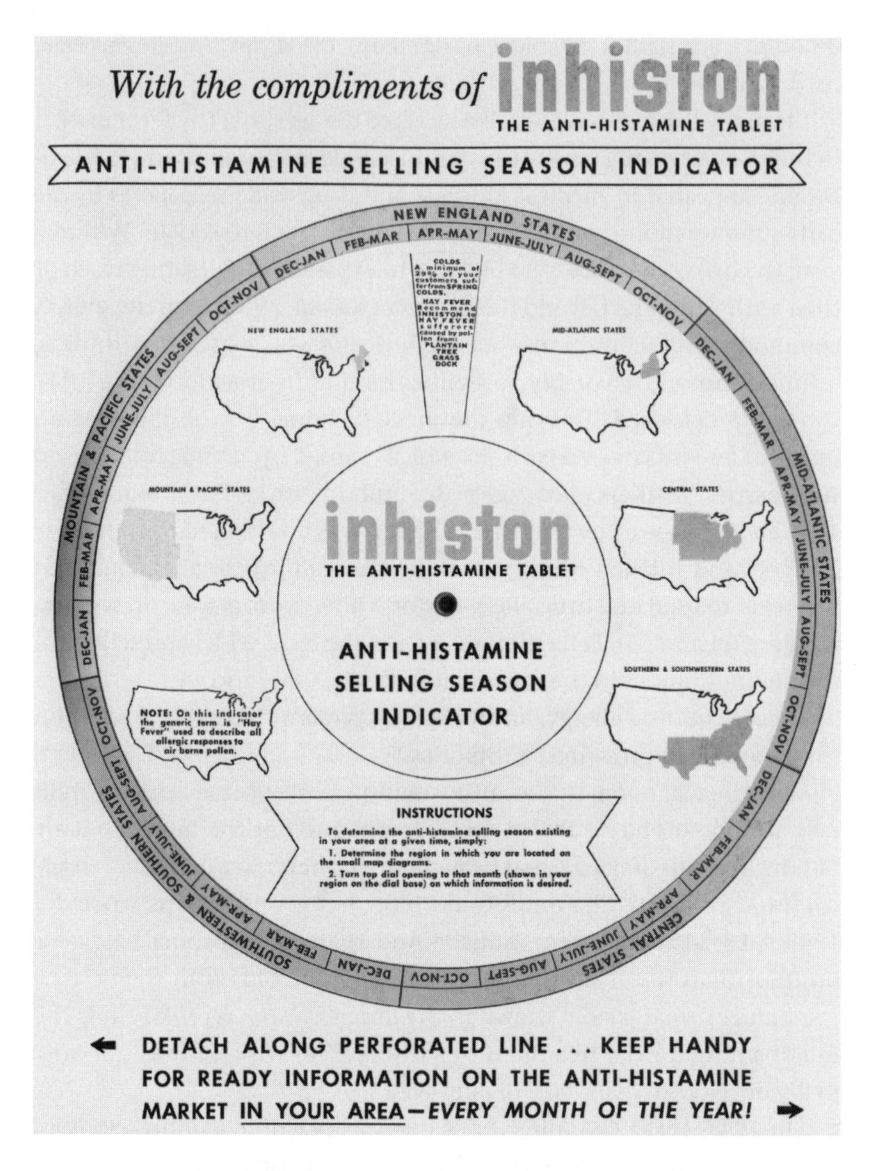

Figure 43. Drug manufacturers pushed to expand the market reach of antihistamines beyond the northeast hay fever season. Pharmacists could consult Inhiston's "Anti-Histamine Selling Season Indicator" to know when they should be promoting antihistamines to consumers in their region.

spasms of the smooth muscles surrounding the bronchial airways that were associated with asthma attacks.[19] In 1947, Searle also introduced its "new antihistamine," Hydrillin, a combination of Benadryl and Aminophyllin. Whether or not the combination constituted a new drug led Searle into a heated debate with the American Medical Association's Council on Pharmacy and Chemistry over fair advertising.[20] But the combination of an antihistamine and an antispasmodic showed Searle's strategy in entering the new market of allergy drugs. Antihistamines such as Benadryl had limited effectiveness in relieving asthma symptoms, and Benadryl in particular could cause severe drowsiness. Gay himself abandoned the clinical use of Benadryl because of the severe side effects experienced by over 60 percent of his patients.[21] By reducing the dosage of Benadryl and combining it with a proven bronchodilator that also had stimulant effects, Searle hoped to minimize the sedative side effects of the new antihistamines while simultaneously capturing the two overlapping markets of hay fever and asthma sufferers with a single drug. It proved a winning combination, not just for allergy sufferers, but also for those afflicted with motion sickness.

One of the experimental drugs that Searle sent to Gray in 1947 was, like Hydrillin, a hay fever/asthma cocktail. It consisted of diphenhydramine, the active antihistamine compound in Benadryl, in combination with a derivative of the bronchodilator drug theophylline. Gay gave samples of the experimental drug to an expectant mother he was treating for hives. The mother-to-be was unimpressed by the drug's effectiveness in controlling her allergic condition under the dose prescribed, but she noticed that her car sickness had remarkably improved since taking the drug. Intrigued, Gay gave her a placebo. Her car sickness returned. Another dose of the real thing and her car sickness disappeared. Gay started dispensing the drug to patients and friends heading on trips, asking them to report back on its effects. While traveling by boat to Europe in the summer of 1948, Gay gave the drug to seasick members of the American Olympic team, who were grateful to have the competitive edge of arriving in London fresh and ready to go.[22]

Impressed by these anecdotal findings, Gay alerted the chief of staff of the U.S. Army of the potential benefit that such a drug might have in military operations. "Operation Seasickness" and the blockbuster drug Dramamine were born. On 27 November 1948, armed with twenty-two thousand white and pink capsules of Dramamine and twelve thousand

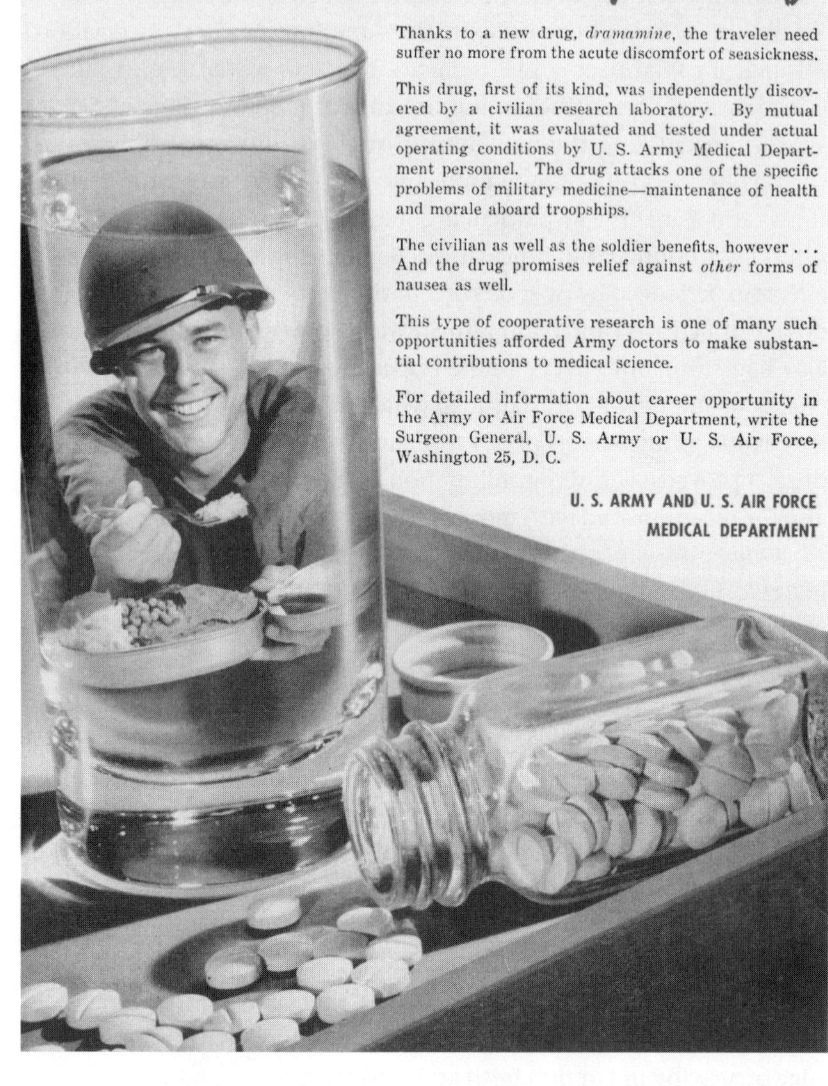

Figure 44. Dramamine was the result of a hay fever/asthma cocktail developed by Searle, clinically tested and proven for the treatment of motion sickness by the U.S. Army on a transport ship to Germany.

white and pink placebos, Gay accompanied 1,376 replacement troops on the U.S. Army transport ship *General C. C. Ballou,* headed for Germany. Weather conditions on the ten-day voyage were extremely rough. On some days, the ship pitched and rolled as much as thirty-five degrees. Within twelve house of departure from New York harbor, Gay reported the "corridors of compartments were congested by sick men, so ill that they were unable to reach the latrines." Of the troops that took Dramamine as a prophylactic, fewer than 2 percent experienced symptoms of seasickness, compared to 28 percent in the control group. Among 539 cases of seasickness, Dramamine dramatically alleviated 94 percent of them.[23]

Gay and his colleague Paul Carliner announced their findings to the Johns Hopkins Medical Society on 4 February 1949. On the same day, the public relations department of both the U.S. Army and the Johns Hopkins University and Hospital released the story of "Operation Seasickness." The Dramamine story took off. Within a matter of days, newspapers, radio stations, and magazines throughout the country were announcing a "cure" for seasickness. Gay was inundated with thousands of letters, cablegrams, and local and long-distance calls from sufferers of seasickness, airsickness, train sickness, and car sickness, all pleading for Dramamine. G. D. Searle rejoiced in the publicity, since as a manufacturer of only ethical pharmaceuticals (prescription drugs), it limited its advertising to physicians and pharmacists. To meet anticipated demand, its plant stepped up production of the drug, scheduled for release on 7 March. Orders from the Army, Air Force, Navy, and the commercial airline industry came pouring in. In August 1949, the Medical Department of the U.S. Army and U.S. Air Force took out a full-page advertisement in *Time* and *Newsweek* promoting its role in the establishment of Dramamine as an effective drug for civilian use. "The Drug that Makes Soldiers Good Sailors Will Soon Be Working for You!" exulted the military ad (Figure 44). Within a year of Dramamine's release, G. D. Searle's stock soared from $38 to $144 per share.[24]

It didn't take long for Searle's competitors to jump on the motion sickness bandwagon. Two weeks after the Food and Drug Administration (FDA) permitted the sale of Dramamine as a prescription drug—for which G. D. Searle alone reaped profits—Dr. A. C. Bratton, director of Pharmacological Research at Parke-Davis, sent a letter to Leslie Gay to ask which component of the Dramamine compound—diphenhydramine or the theophylline derivative—was responsible for the drug's therapeutic effectiveness in motion sickness.[25] Parke-Davis had a huge financial stake in the answer to this question. Searle and Parke-Davis shared the patent on diphenhydramine. To admit that the "active ingredient of Dramamine was diphenhydramine," wrote Irwin Winter, director of Clinical Research at Searle, "would be handing [Parke-Davis] a patent interference on a silver platter." Winter advised Gay to tell Parke-Davis that the activity of the drug "resides in the Dramamine molecule and not in any single component." Given that millions of Benadryl users had yet to notice any effect on motion sickness, neither Winter nor Gay believed that the antihistamine component alone could account for Dramamine's effect.[26] Nevertheless, within a few months after Dramamine's release, the U.S. Air Force, Army, and Navy launched clinical trials in cooperation with Parke-Davis to test Benadryl's effectiveness in the treatment of motion sickness.

From the military's point of view, the key issue in drug trials was performance. For the military, it mattered little if a drug countered motion sickness if at the same time side effects such as drowsiness prevented troops from performing optimally aboard ship or on the battlefield. Marksmanship—how accurately soldiers could fire their guns while under the influence of a test drug—was a decisive consideration in the U.S. Navy's Dramamine trials.[27] Dramamine's low toxicity, minimal side effects, and effectiveness in combating motion sickness gave it the edge over Benadryl, which Gay, Winter, and other clinicians considered to be "too toxic for general use."[28] By 1957, Searle's revenues had grown to $30 million per year, a 300 percent increase in growth since its introduction of Dramamine as the miracle motion sickness remedy.[29]

Whether individuals suffered from hay fever or seasickness, antihistamines enabled them to overcome limits imposed by nature and to work or play at their best. Seasonal pollen, rough seas, high winds, or bumpy terrain posed little threat to a body fortified with the new antihistamines. In 1948 images of people golfing, swimming, ice skating, skiing, and working indoors and out helped sell Schering's first antihistamine, Trimeton, which offered "pleasant relief" to the allergy sufferer through all seasons of the year "without inducing unpleasant reactions" (Figure 45). As an estimated six hundred thousand Americans headed for Europe and another 50 million headed for the beaches and mountains in the summer of 1956, they were encouraged by *New York*

Figure 45. A 1948 ad for Trimeton, Schering's first antihistamine. For over half a century, antihistamines have been sold to American consumers who wish to work and play at their best through all seasons of the year. Reproduced with permission of Schering Corporation. All rights reserved.

Times medical correspondent Howard Rusk to pack Dramamine as insurance for holiday fun.[30] And as troops headed across the ocean for battle, they could relax knowing that antihistamines would allow them to perform aboard ship even in the most severe storms. A simple little pill had freed Americans to get out and live life; they were no longer held back by nature's constraints. Turn on the television and watch any direct-to-consumer antihistamine ad, and you'll still see the same message promoted fifty years later as beautiful people play outdoors, their allergic bodies seemingly oblivious to the environment around them.[31]

In the postwar enthusiasm for science and technology, antihistamines, like the herbicide 2, 4-D, symbolized technological triumph over nature. Physicists had unlocked the secrets of the atom. Biomedicine had uncovered the hidden world of viruses and opened a window onto the secret structure of life. But despite these triumphs, one simple force of nature could still stop the American worker and bring the American economy to its knees: the common cold. Biomedicine had no answer to combat this natural adversary—that is, not until the advent of antihistamines.

If the common cold had gotten "the same amount of brains and planning, and but a fraction of the two billions dollars that went into unlocking the atomic age," exclaimed one physician in 1946, the "age-old mystery could be eradicated."[32] The need to stomp out the common cold seemed especially pressing in the postwar years. Metropolitan Life Insurance calculated that the annual cold season cost the American public over $1 billion per year.[33] One labor spokesperson estimated that "industry had lost five times as much time by the common cold as by strikes." Neither industry nor the military was willing to yield that much down time to nature, particularly when science and medicine had recently harnessed the atom's energies and manufactured miracle antibiotic drugs in the laboratory.[34]

Today, if you step into any drugstore, supermarket, or convenience store in America, you will find OTC allergy relief medications such as Benadryl in the cough and cold section (Figure 46). This near-universal drugstore geography is how most Americans know where to find their preferred cold or allergy medications. But the place of antihistamines on drugstore shelves, and what we expect of them for our bodies, is a legacy of the "Cold War." Fought in the late 1940s and early 1950s, the war was waged by pharmaceutical companies, physicians, the military, and

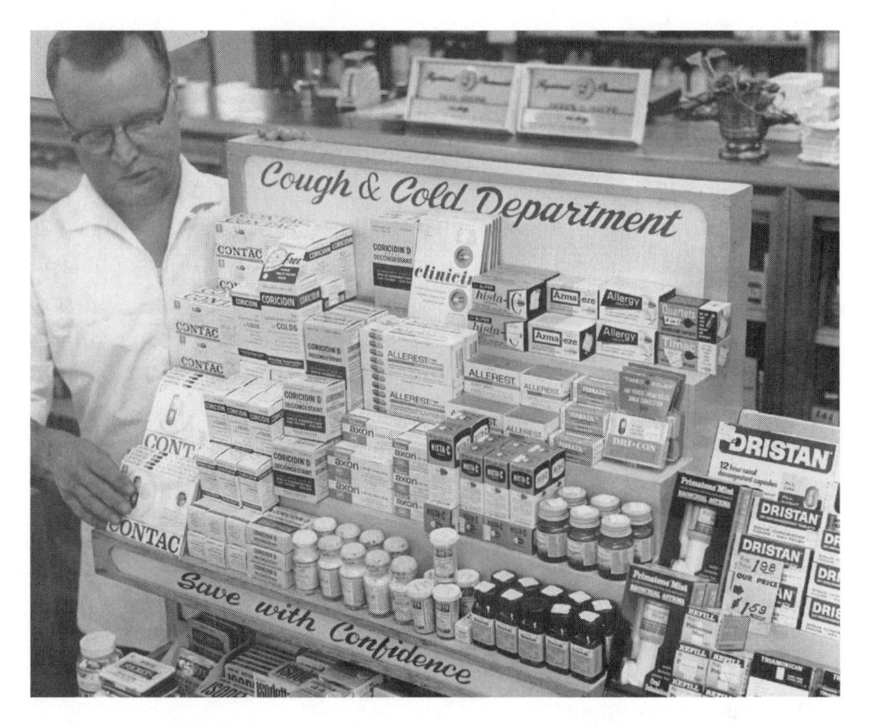

Figure 46. Display of the cough and cold department of Heinz Pharmacy in Creve Coeur, Illinois, in 1969. The inclusion of allergy products such as Allerest and Azma-eze is a legacy of the "Cold War." Courtesy of American Institute of the History of Pharmacy, "Drug Topics Collection."

industry over patients, workplace performance, and the common cold. It also pitted the American Medical Association (AMA) against the proprietary drug industry. Like that of "Operation Seasickness," the Cold War story began with a flexing of the military's biomedical research muscle in the face of nature's adversity.

In 1947, John Brewster, a physician at the U.S. Naval Hospital at Great Lakes, Illinois, reported an anomaly about an allergic patient he had been treating with Benadryl. The patient contracted a cold while taking the allergy medication. But the full-fledged battery of cold symptoms, including sniffling, runny nose, coughing, and congestion, which normally accompanied a bout with the cold virus, never appeared. Suspecting that "many of the common colds have an allergic origin," Brewster began using Benadryl routinely in the treatment of common colds.[35] After good success with more than 100 patients, he designed a clinical trial that included 572 Navy servicemen and their families and involved five different antihistamines. Brewster announced in 1949 in the *U.S. Naval Medical Bulletin* that antihistamine drugs were highly effective in aborting the common cold if taken within an hour or so after the onset of cold symptoms.[36] In other American workplaces—including the Dennison Manufacturing Company, Sylvania Electric Products Corporation, DuPont Corporation, and Sing Sing Prison—similar reports began to appear of successful experiments with the administration of antihistamines to employees as a preventative treatment against the common cold. The DuPont plastics factory in Arlington, New Jersey, noted a 60 percent drop in man-hours lost when 1,000 employees were given a daily dose of Hydrillin, compared to previous years without worker medication.[37]

To some clinicians, these findings seemed to suggest an allergic basis for the common cold. How else could one account for antihistamine's miraculous powers in cold relief? Perhaps, as an editorial in the *Journal of the American Medical Association* speculated, the common cold was in fact an "allergic response . . . to contact with a specific protein, which is the cold virus or its products."[38]

Hay fever had a long history of association with seasonal colds, evident in such nineteenth-century names for it as rose cold, hay cold, and autumnal catarrh. By the early twentieth century, however, the clinical field of allergy firmly differentiated between these two illnesses. Physicians advised patients not to confuse seasonal allergies with the common cold and urged patients to consult a doctor if their cold symptoms appeared on a regular or seasonal basis. Allergy, unlike the common cold, doctors stressed, required the care and advice of a specially trained physician. But now, in a prestigious medical journal, "catching cold" became linked to allergy and its treatment. So long as that treatment—antihistamines—remained available only by prescription, cold sufferers seeking antihistamines would significantly increase the number of bodies entering doctors' offices.

What physicians feared and what provoked the most alarm was the prospect that antihistamines would become OTC drugs. If this happened, physicians who prescribed them would lose the large cold market and, perhaps, some of the allergy market as well. Not only cold sufferers, but allergy patients too would simply self-medicate with nonprescription antihistamines. In battling the common cold, Brewster favored making antihistamines readily available to the public. "Real progress in the elimination of the common cold," he argued, "will be made when and if an antihistamine drug which is effective and yet sufficiently safe to permit the public to purchase it without a prescription is developed."[39] Within weeks, what physicians had most feared happened. Drug companies began filing applications with the FDA to permit OTC sales of antihistamines as cold remedies. The Cold War was under way (Figure 47).

On 2 September 1949, the FDA cleared Neohetramine as the first antihistamine for sale to the public without a prescription. Its manufacturer, Nepera Chemical Company, also received permission in early October for OTC sales of Anahist, an identical product manufactured by Nepera's subsidiary, Anahist Company. The Schering Corporation, a subsidiary of Union Pharmaceutical Company, also entered the Cold War in October with Inhiston. In December, famed science writer Paul De Kruif announced "the beginning of a great step ahead in the comfort of mankind" in his *Reader's Digest* article, "Is This at Last Good-Bye to the Common Cold?"[40] Anahist timed the release of De Kruif's article with its own marketing blitz: ads on coast-to-coast radio programs; in *Good Housekeeping, Look,* and other national magazines; and in display ads in over sixty-three newspapers. Sales of Anahist soared from nothing to $1 million in a single month. The whole American economy was profiting from the Cold War boom. Schering was shipping out 30 million Inhiston tablets a week. Manufacturers of carton, bottle, and other drug-packaging products couldn't keep up with the demand. One cap supplier reported producing more than one hundred thousand pill bottle covers in a single night. Inhiston even introduced an antihistamine dispenser unit, the Cold Control Plan, for companies to place right above the factory water cooler so that employees might have easy access to the drug. Under the influence of antihistamines, workers kept working and consumers kept buying. It was an American dream.[41]

But soon the bubble burst. The initial attack on OTC antihistamine drugs came not from consumers or the FDA but from the AMA. In the same month that Paul De Kruif heralded antihistamines as the new miracle drugs for the common cold, Austin Smith, editor of the *Journal of the American Medical Association,* declared that "no one yet knows what harmful effects [antihistamines] may produce on the body in general, or on specific tissues, when taken over prolonged periods of time." In New

Winning the COLD WAR!

THE ANTIHISTAMINE THAT'S MAKING SALES HISTORY!

Leading Chain Store Executive Says:

"*ANAHIST is the biggest traffic builder we ever had . . . and it gives us good profit margins, too!*"

Retail F. T. M.'s Large—99¢ Medium—55¢

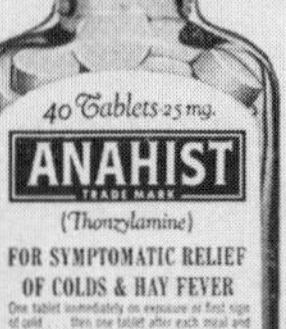

THE ANTIHISTAMINE DESCRIBED IN READER'S DIGEST!

"Here's the best health news of the year," says Reader's Digest, December issue.

Telling millions of people about the new antihistamine proved clinically-effective for colds, the article concludes, "*The public will purchase it under the trade name ANAHIST.*"

Facts to Remember about Antihistamines:

Safety . . . Because antihistamines differ as chemical compounds, they vary greatly in performance and effectiveness. "Because of its low toxicity at therapeutically effective doses, ANAHIST is recommended as the safest antihistamine drug." (Long Island College of Medicine)

Effectiveness . . . ANAHIST's 25 mg. antihistamine tablets provide the CLINICALLY-PROVED EFFECTIVE dosage for colds, now available without a prescription. Remember, laboratory tests show that ANAHIST's 25 mg. tablets will combat 2½ times as much destructive histamine in the system as any 10 mg. tablet on the market! You can sell ANAHIST with confidence—it is the antihistamine recommended for family use!

ANAHIST sells on sight—quickly, profitably! Attractive, eye-catching counter display carton free with order! Phone your wholesaler now!

Powerful Sales-Making National Advertising!

Magazines . . . Full page ads in LOOK, COLLIER'S PARENTS' and other leading national magazines!

Newspapers . . . Big space ads in the country's biggest daily newspapers! *PLUS* NANCY SASSER!

Sunday Supplements . . . Full page ads in PARADE, AMERICAN WEEKLY, THIS WEEK and others.

. . . Plus 2 New Radio Shows! ANAHIST now sponsors *two* radio programs coast to coast! Families everywhere listen to "*True or False*" on Saturday, "*The Falcon*" on Sunday!

ANAHIST CO., INC., Yonkers 2, N. Y.

©1950

CALL YOUR WHOLESALER FOR ANAHIST NOW!

TRADE MARK

Figure 47. "Winning the Cold War!" *American Druggist,* January 1950, 49.

Jersey, physicians of the Middlesex County Medical Society passed the first resolution calling upon the FDA to "publicize the inherent dangers of the indiscriminate use of these drugs." The *New England Journal of Medicine* also lashed out. "This is the most striking example to date," the journal's editor exclaimed, "of the advertising methods of manufacturers and promoters who are steadily going over the heads of the medical profession in attempts essentially to force physicians through their trumped-up public demand to accept remedies before their usefulness has been adequately substantiated and ill effects determined."[42]

Questions of the efficacy of antihistamines in treating and preventing colds first came to public light in February 1950 in a report issued by the Council on Pharmacy and Chemistry of the AMA. The report concluded that no "scientifically acceptable study" had yet been performed to "test the true effectiveness of the antihistaminic drugs in the control of the common cold."[43] The council was particularly critical of Brewster's original study. Questioning the study's experimental design, it noted that many of the beneficial effects Brewster had observed in his subjects could have been due to the alleviation of symptoms associated with an allergic condition rather than the common cold. Brewster had relied upon the self-diagnosis of patients, and many of his subjects, the council reasoned, might be confusing a cold with an allergic condition. This seemed particularly plausible, argued the council, given that those who sought early treatment with antihistamines in Brewster's study were those for whom cold-like symptoms had become "a serious illness or a threat to their jobs or their social plans."[44] A few months later, in a clinical trial that included almost two thousand people, the Naval Medical Research Unit at the Great Lakes Naval Training Station issued a press release stating that "no significant difference" could be found between the use of antihistaminic drugs, typical cold tablets, or placebos in the prevention of the common cold.[45]

These new findings set in motion an investigation by the Federal Trade Commission (FTC) in the spring of 1950 into whether drug manufacturers, in touting their products as a "cure" for the common cold, had violated federal laws relating to interstate commerce and false advertising. In 1950, drug companies had only to prove to the FDA that their products were safe for OTC use when taken as directed on the product's label. Claims of a drug's effectiveness were handled by the FTC.

While the FTC deliberated over whether drug companies had violated federal trade laws, physicians stepped up their attack by alerting the public to the inherent dangers of antihistamines, particularly for children. In an alarming article in the *Ladies' Home Journal,* Herman Bundesen, president of the Chicago Board of Health, warned parents of the tragedies that might befall their families from indiscriminate antihistamine use. Bundesen reported on three recent cases of young children who had either died or lapsed into unconsciousness after ingesting OTC antihistamine drugs. Children taking antihistamines for allergy or colds, advised Bundesen, needed to be placed under a physician's care. Bundesen also advised caution and restraint in embracing antihistamines as the magical cure for children's allergies. As much as parents might wish it, antihistamines did not eliminate the need for shots, special diets, or environmental control of the home. The therapeutic effectiveness of antihistamines varied, Bundesen insisted, according to place—in this case, the body of an individual child.[46]

In June 1950, the FTC reached an agreement with five pharmaceutical companies. The agreement permitted the continuing sale of OTC antihistamines provided that the companies did not advertise that their products would "cure, prevent, abort, eliminate, stop, or shorten the duration of the common cold." Claims that their products stopped cold symptoms were allowed. The ruling had little effect on drug manufacturers. A spokesperson for Anahist told *Drug Trade News* that after a careful review of its advertising, the company was continuing its 1950 summer and fall advertising schedule as planned. In fact, Anahist advertisements and packaging had promoted the product only for "symptomatic relief of colds and hay fever." The drug industry also interpreted the FTC settlement as a federal "clean bill of health" for its products when used as directed. But the settlement said nothing about safety. Safety concerns fell under the jurisdiction of the FDA.[47]

The AMA fought a losing battle in its effort to limit the sale of antihistamines to the public. To some, the AMA's actions appeared as a noble call to safeguard public health against an industry exploiting the pocketbooks and bodies of American consumers. To others, the actions appeared more cynical. In April 1950, an editorial in *Collier's* suggested that the "real purpose of the AMA objection" to the OTC sale of antihistamine drugs was "to give the prescribing physician a monopoly on the first promising remedy that ever came along."[48]

The opening Cold War battle was but the beginning of an ongoing fight among physicians, drug companies, federal agencies, and consumers over the control of allergic bodies in America. Would antihistamines and other allergic drugs make treatment of allergic illnesses such as hay fever and asthma more akin to that of the common cold, which "for all practical purposes" doctors had abandoned to the public?[49] Or would such drugs necessitate that allergic Americans relinquish some control of their bodies to physicians specializing in allergic disease?

Despite the warnings of medical organizations such as the AMA, the market in antihistamine drugs continued at a brisk pace throughout the 1950s. Of prescribed antihistamines, a good share was being prescribed by specialist physicians.[50] By the 1960s, the antihistamine prescription market had pretty much stabilized at roughly 39 million prescriptions filled each year. In dollars, those prescriptions accounted for approximately $42 million in annual sales.[51] Sales figures for OTC antihistamine products are not available. While the debate over OTC versus prescription allergy drugs would flare up again in the 1980s, at a time when the number of patients seeing clinical allergists was declining, one thing was certain: antihistamines had taken a place in American culture as remedies for the symptoms of allergy and the common cold. As Nathan Wishnefsky, a drug market analyst, astutely observed in the 1950s, whether antihistamines cured "the virus type of cold or the allergic type of cold" was of "academic interest." All that mattered to the American consumer was if such drugs permitted him to "remain at work" and "continue uninterruptedly both his daily task and his social pleasures," immune to the ill winds and seasonal whims of nature.[52]

For a while, antihistamines seemed to be the magic bullet that might ameliorate the symptoms of all allergic diseases—hay fever, food allergies, hives, and asthma. But the hopes of creating a universal miracle drug that would be effective regardless of the individual patient or place soon vanished. Although companies initially marketed antihistamines such as Benadryl and Pyribenzamine to treat a whole host of allergic conditions, by the mid-1950s it was clear that antihistamines offered little relief against asthma. They dried up runny noses and soothed itchy skin, but they did not relax the airways of asthmatic patients struggling to breathe. Bronchodilators did.

Since the early 1900s, epinephrine and ephedrine were the physician's drugs of choice in helping relieve an asthma attack. Parke-Davis chemists isolated epinephrine in 1901 and trademarked it under the name Adrenalin. Ephedrine was derived from the *ma huang* plant in China, where for centuries the herb had been used as a traditional remedy in the treatment of asthma and other respiratory problems. By the 1930s, both epinephrine and ephedrine were administered by either injection or a hand-held nebulizer that delivered an exceedingly fine spray of epinephrine or ephedrine solution into the lungs. During an asthma attack, the smooth muscles surrounding the bronchioles, or small airways of the lung, go into spasms triggered by some type of irritant—pollen, house dust, mold, or ozone (among others). The adrenergic bronchodilators act by relaxing the bronchial muscles. Their actions are not limited to the lungs, however. The body, like any ecosystem, is a complex web of interrelationships, and once these chemicals are in the lungs, they move to other parts of the body. Effects on the heart can result in palpitations, dizziness, and irregular heart rhythm. Action on the nervous system can cause side effects such as anxiety, tremors, headache, or nervousness. By the 1930s, aminophylline and theophylline had also become available. While the biochemical pathway of their action is slightly different from epinephrine or ephedrine, their therapeutic effect is the same: they too relax the bronchospasms associated with an asthma attack.[53]

Although the wartime chemical industry did not revolutionize bronchodilator drugs, it did initiate a change in packaging that transformed the technology of drug delivery: aerosols. During World War II, research scientists working at the U.S. Department of Agriculture (USDA) developed a new, convenient means for servicemen to combat disease-carrying insects that threatened the war effort. They invented a small aerosol can that utilized Freon, DuPont's fluorinated hydrocarbon, as a pressurized propellant to dispense potent pesticides such as DDT. When the war ended, Bug Bomb, Slug-A-Bug, Moth Proofer, and other such products quickly found their way onto retailers' shelves.[54] Consumer demand for new aerosol items was growing so quickly that DuPont ran out of Freon supplies in the fall of 1953. In 1955, insecticides, room deodorants, shaving lather, hair lacquer, and Christmas tree snow dominated the list of best-selling aerosols.[55] Drugstores stocked spray products more quickly than all other retail outlets. Given the lead drugstores had over other retailers in the aerosol trade, it took little time for drug companies to find ways to package their chemicals in handy, easy-to-use sprays.

Pressurized aerosol canisters went from delivering insecticides to the front lines to delivering drugs to the lungs of wheezing asthmatics. In 1956 the first two self-propelled, metered-dose aerosolized medications hit the market: Medihaler-Epi and Medihaler-Iso. Manufactured by Riker, Medihaler-Epi shot a 0.15 mg dose of the bronchodilator epinephrine, in addition to fluorinated hydrocarbons. Medihaler-Iso delivered a 0.06 mg dose of a new bronchodilator drug, isoproterenol sulfate.[56] Simple to use, easy to carry, and fast-acting, metered-dose bronchodilator inhalers quickly became the rage as both prescribed and OTC asthma drugs. During the 1960s, their rate of growth far exceeded that of antihistamines in the allergy drug trade. With over $37 million in sales in 1970, prescribed bronchodilators had surpassed the earnings of antihistamines by over $1 million. Asthma, once a small, specialized niche market for the drug industry, was starting to look like a field for lucrative growth and market expansion.[57]

Despite drug improvements in the treatment of asthma—in addition to bronchodilator aerosols, corticosteroids helped reduce airway inflammation in severe asthma cases—physicians in the 1960s began to note an alarming trend. In Australia, England, New Zealand, Norway, and Wales deaths from asthma were on the rise. In England and Wales, the number of asthma deaths had nearly doubled between 1959 and 1966. The problem was especially visible among children between the ages of ten and fourteen. In this age group, an eightfold increase in the asthma mortality rate had occurred in six years.[58] The skyrocketing increase in asthma deaths in these countries prompted a heated debate in medical journals and led to numerous epidemiological studies to determine the cause. Could the trend simply be a reflection of changing diagnostic categories of asthma—that is, a statistical artifact of how asthma deaths were classified? Or might it be a result of changing environmental conditions, particularly increasing air pollution? Or perhaps new methods of treatment—notably corticosteroids and inhaled bronchodilators—were responsible for the trend?[59]

By the late 1960s, medical opinion considered the use of bronchodilator inhalers as the most likely explanation for the marked increase in asthma deaths. While corticosteroids had been in use in Great Britain since 1952, pressurized aerosol inhalers first became available in England and Wales in 1960. Over the next five years, their sales increased fourfold. Inhalers such as Medihaler-Iso were those most commonly found in the medicine cabinets of patients who died. Of the patients who died, 86 percent had inhaled bronchodilators within the last month of their life. Correlation, however, does not prove causation. Researchers cited many reasons why inhaler use might be a factor in the rise in asthma deaths. One explanation suggested that the deaths simply reflected cases of patients with severe asthma who had relied on self-medication with OTC bronchodilators rather than seeking medical attention. Other hypotheses focused on the side effects of chemicals in the freon-propelled, metered-dose inhalers. Once in the circulatory system, isoproterenol (the active ingredient in Medihaler-Iso) might readily induce abnormal heart rhythms in an organ already sensitized by low levels of arterial oxygen or by the fluorinated hydrocarbons used as propellants. Alternatively, excessive use of bronchodilators might cause a rebound effect by exacerbating rather than reducing the constriction of the airways. Whatever the reason, Britain's Committee on Safety of Drugs believed the evidence warranted a ban on direct sales to the public of bronchodilator inhalers in 1968. And asthma deaths promptly declined.[60] Paul Stolley, an epidemiologist and physician at Johns Hopkins School of Hygiene and Public Health, called it "the worst therapeutic drug disaster on record. There's nothing else—not even thalidomide," he exclaimed, "that ranks with it."[61] (Thalidomide was a drug widely prescribed in Germany and Britain during the 1950s and 1960s to pregnant women with morning sickness. It resulted in birth defects in more than twelve thousand children.)

Stolley became curious about the mysterious epidemic of asthma deaths while he was on vacation in England. During his stay in a small English town, he was asked by villagers to see a twenty-four-year-old asthma patient who had succumbed to a severe attack in church. Stolley went to the patient's home, only to find him dead, an inhaler clutched in his hand. After further investigation, the American doctor found that the particular bronchodilator marketed in Britain, Australia, New Zealand, and Norway was made by an American pharmaceutical firm and delivered five times the dose of isoproterenol per spray as the comparable product sold in the United States. Stolley wasn't willing to indict the American company's inhaler as the sole cause of the increase in asthma deaths in England, but he suggested that it was a contributing factor of the epidemic's severity and a reason why the United States had been spared.[62]

The bronchodilator scare abroad prompted American physicians in the late 1960s to reflect critically on the therapeutic use of inhalers in asthma treatment in the United States. As the sales of inhalers jumped from 3.3 million in 1964 to 5.1 million in 1968 in the United States, disturbing reports of patient abuse began to appear. Many asthmatic patients had become addicted to their bronchodilator inhalers, often grasping them for dear life. Some patients admitted to going through an entire inhaler in one day. One woman with severe asthma, who displayed extreme sensitivity to house dust and tobacco smoke (among other allergens), was living with two very heavy smokers and using an isoproterenol inhaler every few minutes. As these reports suggested, it was far easier to shoot a dose of a drug into one's lungs than to address environmental triggers, some within one's control, others perhaps not, that exacerbated asthma. And since the interior spaces of the body were not readily perceived, it was relatively easy to ignore possible side effects. Out of sight, out of mind.[63]

The abuse of bronchodilators prompted American physicians to urge caution in treating asthma through drugs alone. In 1969, the Allergy Committee of the American College of Chest Physicians voted to urge the FDA to limit the availability of OTC bronchodilator drugs. Possible harmful effects also led to greater caution among physicians in prescribing metered-dose inhalers. Within a year after Stolley's study, refill prescriptions of Riker's Medihaler-Iso plummeted by more than 40 percent.[64]

Drugs alone were not enough in the care and treatment of asthma. One year before the initiation of Earth Day in 1970, Murray Dworetzky advised his fellow members of the American Academy of Allergy to become "intimately concerned with the total environments of [their] patients"; these included, "in addition to physical and chemical factors, social, economic, psychological, and educational factors as well." The future of the asthma patient, Dworetzky suggested, depended upon the recognition of asthma as a "multicausal disease." Physicians, he argued, needed to expand their vision beyond symptoms and drugs to attend to the patient as a whole, including the environment in which he or she lived. Asthma patients needed a comprehensive, integrated approach to medical care.[65]

At the very moment allergists in America called for a more integrated, ecological approach to asthma management and care, breakthrough discoveries were being made in the immunology of allergic disease. These discoveries fueled faith in new and better drugs to come—and profits to be made. It was a mindset that kept the focus of attention in asthma treatment on the chemical alteration of the body. As noted above, in 1967, Kimishige and Teruko Ishizaka isolated a unique immunoglobulin, found in association with antibodies in the sera of hay fever patients. That same year in Sweden, Gunnar Johanson and his colleague Hans Bennich discovered a similar immunoglobulin in the serum of a patient with an unusual myeloma and also found its presence in the sera of patients with allergic asthma. In 1968, the World Health Organization officially named the protein immunoglobulin E, or IgE.[66]

After more than forty years, biomedical technology had made visible the mysterious active substance that rendered Carl Prausnitz's body sensitive to fish when he injected himself with the serum of his allergic colleague and broke out in hives after eating a seafood dinner. IgE's presence in heightened levels in allergic bodies suggested that it held the biochemical, perhaps even the genetic, clue to allergic disease. IgE became the holy grail of allergy, the molecular key that might unlock the biochemical mystery of the allergic body. Histamine was far down the chain of the allergic reaction. IgE brought biomedical researchers much closer to the source.

Over the next two decades, scientists began to paint a detailed picture of the immune landscape and the biochemical reactions involved in an allergic reaction. By the early 1980s, a whole new world had become visible inside the body. When an allergen entered an allergic body, the body made IgE specific to the allergen. In this way, the allergic body became sensitized. IgE molecules, circulating in the bloodstream of allergic bodies, attached themselves to mast cells residing in tissue and basophils harbored in the blood. (Both mast cells and basophils come from the bone marrow and are a part of the immune system.) When the allergen next made its way inside the body, IgE molecules bound themselves to the offending substance. When two adjacent IgE molecules on the surface of a mast cell were connected by an allergen, it unleashed a flood of events that altered the cell's permeability and activated a series of enzyme reactions that caused mast and basophil cells to release their contents. Histamine, prostaglandins, leukotrienes, and cytokines were let loose into the surrounding tissue and bloodstream. These were just a few of the chemical mediators that over time became visible to biomedical researchers and that accounted for many of the symptoms experienced by allergy sufferers.[67]

Although it was two decades after the discovery of IgE before the visible details of the immune landscape yielded new drug therapies, the market for asthma drugs in the United States continued to grow through the 1970s and 1980s. Between 1972 and 1985, prescription drugs overall in the United States showed a modest 7 percent increase in the number of prescriptions filled. During the same period, prescriptions for asthma drugs increased 200 percent. The asthma market was booming, and it owed its phenomenal success largely to the introduction of two products in the late 1970s: beta agonist inhalers and slow-release (or sustained-release) theophylline. Beta agonist inhalers were a variation of the older class of inhaled bronchodilators that had sparked so much debate during the previous decade. The earlier bronchodilators, which relied on epinephrine and isoproterenol, acted not just on the lungs but on multiple organs in the body and were particularly notorious for stimulating the heart. The new beta agonist drugs, which included albuterol, metaproterenol, and terbulatine, acted selectively on the smooth muscles of the bronchioles with minimal stimulating effects on the heart. From 1977 to 1985, beta agonist inhalers doubled their share of the asthma market, accounting for 25 percent of all prescriptions dispensed for the treatment of asthma by 1985. Sustained-release theophylline, a longer-acting preparation of an oral bronchodilator drug in use since the 1930s, took over an even greater share of the asthma drug trade. In 1985, immediate-release and sustained-release versions of this drug made up nearly 50 percent of all prescriptions written for asthma drugs.[68]

Allergy was a fast-growing market for pharmaceutical companies in the 1980s. One formidable power in the market's rapid expansion was Schering-Plough, a leader in allergy, asthma, and cold products since its introduction of Inhiston in the "Cold War" years. In 1981, the company launched Proventil, a beta agonist inhaler with albuterol as its active ingredient. Proventil quickly became the company's best-selling product; in the 1980s revenues from Proventil grew at a rate of 40 percent or more each year. By the time Schering's patent on Proventil expired in 1989, annual sales of the drug had surpassed well over $200 million, and it had captured one-third of the entire beta agonist bronchodilator market.[69]

Beta agonist inhalers such as Proventil opened not only the airways of asthmatics; they also opened new disease categories, as well as access to new patients. Bronchial spasms that could be relieved by the inhalation of a drug like albuterol became a common diagnostic criterion of asthma during the 1980s. The definition of asthma as a disease became intimately linked to response to the preferred drug for treatment. So long as inhaled bronchodilators ruled as the drug of choice, asthma was largely understood as chronic, reversible airway obstruction. As the sale of inhalers like Proventil climbed, a new type of asthma was identified: exercise-induced asthma.[70]

When I was growing up, running on a track team was the furthest thought from my mind. It didn't take a doctor to tell me that aerobic exercise could provoke an asthma attack. I dreaded gym class. It wasn't because I didn't like physical activity. It was the competition I hated, knowing that I would inevitably come in last in any race. I channeled my energies into other outdoor activities and sports—backpacking, rock climbing, martial arts—where the pressures seemed less than on the track field, tennis court, or basketball floor.

But the rapid expansion of the asthma drug market in the 1980s had a curious effect. Running suddenly became a marketing tool and a measure of success in the treatment of asthma. Athletes became the new salespeople for asthma drugs (Figure 48). They also became a new market. One year after Proventil hit the U.S. market, Schering produced a promotional film, *Running Hard, Breathing Easy.* The film featured UCLA track star Jeanette Bolden, the record holder for the collegiate women's indoor fifty-yard dash. Bolden didn't start out as a born athlete. Instead, she grew up in Los Angeles as a sheltered African American child with severe asthma. At the age of thirteen, she was placed in the Sunair Home for Asthmatic Children, a residential treatment facility. There she gained control of her asthma and also learned she could be as competitive as other kids. In high school, she began winning medals in track and field. But in 1978, when Bolden was eighteen, her track career started on a downward spiral. She was hospitalized for asthma. Bolden volunteered

Figure 48. During the 1980s, as sales of bronchodilators soared, athletes became important consumers and promoters of asthma drugs. Photo from Schering-Plough's 1981 annual report. Reproduced with permission of Schering Corporation. All rights reserved.

to take part in Schering's clinical trials of albuterol, and the film shows her using a Proventil inhaler. The newer drugs, her doctor informs us, made a difference in Bolden's life. Four years after her hospitalization, Bolden led the UCLA Bruins to their first outdoor NCAA track-and-field championship by placing second in the hundred meter and running on the Bruins' four-hundred-meter relay team, which finished third in the competition.[71]

Bolden wasn't the only American athlete whose performance had been greatly improved through asthma drugs. An examination (through spirometry and drug technologies) of athletes on the 1984 U.S. Summer Olympics team, which competed in Los Angeles, where Bolden earned a gold medal on the U.S. four-hundred-meter relay team, showed that 67 out of 597 had asthma. Ironically many of these world-class performers did not know they had the disease. What's more, the 11 percent of the team members diagnosed as asthmatic won 13 percent of the medals. The asthmatic athletes were, researchers told *Science News,* victims of exercise-induced asthma, which causes coughing, chest tightness, and wheezing shortly after heavy physical activity. Suddenly articles started appearing in popular magazines, such as *World Tennis,* alerting recreational athletes to be on the watch for the "onset of wheezing with a cough and shortness of breath during physical exertion." Cold weather, noted *World Tennis,* could be a major trigger in provoking an exercise-induced attack in the nonallergic athlete. Articles on exercise-induced asthma steered athletes to the same drugs that worked for allergic asthma: theophylline and beta agonist inhalers. A new disorder and a new market were made. With the right drugs, athletes suffering from either allergic or exercise-induced asthma were no longer limited by weather, seasonal pollens, or their own bodies, Dr. William Pearson told *Science News* in 1984, but "only by their desire and will to succeed."[72]

Bodies, drugs, markets, and medicine were getting ever more tightly linked in the world of allergy and, more generally, health care as the 1980s passed. Allergists had by and large kept a healthy distance from the pharmaceutical industry in the immediate postwar era of drug development. In 1951, the highlight of the American Academy of Allergy meetings was a cocktail party sponsored by the Schering Corporation; as the academy's president observed, the party added "sociability" to a dry scientific meeting.[73] Little more than cocktails flowed at these meetings.

And when the American Academy of Allergy's Drug Evaluation Committee found products such as R. J. Strasenburgh's experimental antihistamine Histionex ineffective in its clinical trials, drug companies made no signs of holding a grudge. Edwin Hays, Strasenburgh's research director, still sent a check for $2,000 to the academy's research fund, thanked the academy for its work, and was "convinced that the evaluation was sound and accept[ed] its implications."[74] In the 1980s, all that professional decorum changed quite dramatically as pharmaceutical companies sniffed, not just pollen, but money, in allergy.

Marketing drugs and marketing a medical specialty became synonymous as disease, drugs, and doctors became more closely wedded than ever before. In 1985, Bruce Simpson, vice president of sales and marketing at Fisons Corporation, an important player in the allergy drug industry, alerted executive officers of the AAAI to data that indicated a declining trend in patients visiting the offices of board-certified allergists. Prescription drug—or ethical pharmaceutical—companies such as Fisons were also concerned that their considerable profit margins were threatened. The threat came from both an increasing trend toward self-care and an FDA push to make more allergy and asthma medications available over the counter. On 15 December 1985, Simpson met with representatives from the AAAI, the American College of Allergy, and the American Association for Clinical Immunology and Allergy. Their purpose was to discuss and finalize "A Joint Statement on Marketing the Allergist . . . Now and for the Future." Out of this meeting the Allergy Marketing Task Force was created to educate the public about allergies and to sell the consumer and third-party payers on the value of the professional allergist and the drug industry in the treatment of allergic disease. For a contribution of $12,500 to the task force, a drug company would get a "gold card" that would grant its representatives privileged access to physicians. For $25,000, a drug company was guaranteed a place on the Allergy Marketing Task Force board.[75]

In the spring of 1986, the task force hired Public Communications, a prominent Chicago-based public relations and marketing firm, to launch an allergy campaign through the national media at a cost of $250,000. The marketing blitz was carefully timed with National Asthma and Allergy Awareness Week and with the spring allergy season. Television spots on allergy appeared on *ABC World News Tonight* and *Good Morning America*. Informational articles that highlighted the extent of allergies—

affecting 35–40 million Americans—and the costs to the economy—estimated at $1 billion—blanketed national newspapers such as *USA Today* and more than 68 regional markets. Radio interviews with carefully chosen and trained allergy experts appeared on more than 589 radio stations across the country.[76] Although many of these articles and interviews did a good job of educating the public about allergic disease and the triggers that can provoke hay fever or an asthma attack, they also had a common refrain. As Dr. Donald Aaronson, a Des Plaines, Illinois, allergist, told readers of the *Chicago Tribune,* if OTC antihistamines didn't work for them, they should consult an allergist. "There is no longer any need for a person to suffer," remarked Aaronson. "Loads of new drugs have been developed within the past three or four years and they give most people relief from allergic symptoms."[77] Educating the public about allergic disease was also an advertisement for the pharmaceutical industry, which, along with the AAAI and other medical specialty groups, helped to bankroll the allergist marketing campaign.

Would consumers seek health care at the cough, cold, and allergy section of the drugstore? Or would they visit an allergist's office, where they could get prescriptions for the latest, more expensive allergy and asthma drugs? As relationships between physicians and drug companies grew closer and more financially intertwined, the AAAI found itself being heavily lobbied from all sides of the pharmaceutical industry. Through its position statements, subcommittees, and representation on FDA advisory panels, the AAAI could greatly affect a drug company's earnings. Ethical drug companies that specialized in prescription drugs and cutting-edge therapies vied for influence over proprietary drug companies, which had traditionally catered to the OTC and generic drug trade. But the once clear distinction between ethical and proprietary drug companies blurred in the prosperous 1980s, when the sales of pharmaceutical companies began their dizzying climb to what is now an annual $200 billion market in prescription drugs.[78] How to remain above the fray of influence peddling by the drug industry posed a great challenge to medical societies like the AAAI.

Navigating a path through an entangled web of bodies, consumers, drugs, and corporate interests wasn't easy. How difficult the path was became plainly visible in 1985. That summer, drug manufacturer Boehringer Ingelheim approached the AAAI to set up an ad hoc committee to make a recommendation on the safety of Alupent, its metaproterenol sulfate inhaler, for OTC use. Alupent MDI, one of the new beta agonist inhalers, had been in use as a prescription drug in the United States since 1973. From 1973 to 1985, an estimated 6 billion doses of metaproterenol sulfate had been consumed by asthmatic Americans, while revenues from the sale of 20 million inhalers helped sustain the U.S. economy as well as Boehringer Ingelheim.[79] In the 1970s, growing public skepticism about the health care industry and rising inflation in health care costs spurred the FDA to take action. It began a systematic review of prescription drugs that might merit OTC status in an effort to drive down health care costs. Metaproterenol sulfate was one such drug. In 1982, the FDA recommended OTC status for metaproterenol sulfate inhalers—specifically Alupent and Metaprel, made by Dorsey Laboratories. Basing its decision on available evidence, the FDA argued that Alupent and Metaprel were as safe as and "more effective than currently available OTC epinephrine [inhalers]" such as Primatene Mist.[80]

In March 1983, Boehringer Ingelheim launched an intensive television advertising campaign notifying asthma patients that "prescription strength" Alupent was now available without the need to see a doctor. The AAAI, the American College of Allergy, and other medical organizations and clinicians were up in arms. England's bronchodilator scare of the 1960s came to the forefront of doctors' minds, but an awareness of lost market share in allergic bodies no doubt did as well.[81] James Mann, an FDA physician, told the *New York Times* that such critics "were simply opposed to 'the whole concept of over-the-counter marketing' with its emphasis on self-medication by patients."[82] But when the threat of a congressional inquiry by the House Subcommittee for Oversights and Investigations appeared, the FDA rescinded its ruling. After less than three months, Alupent and Metaprel were removed from the cough and cold sections of drugstore shelves and were back on the pharmacist's shelf as prescription-only drugs.

Boehringer Ingelheim was not about to back down. Its patent would expire in 1986, at which time other companies would be able to manufacture metaproterenol as a generic prescription or an OTC drug. Boehringer Ingelheim desperately wanted to gain a competitive edge by establishing brand loyalty for Alupent before other companies could start competing. In June 1985, it approached the FDA to reconsider its ruling regarding the OTC status of Alupent. The FDA agreed but only on the condition that Boehringer Ingelheim had the support of medical groups.

It was, after all, opposition from groups like the AAAI that had led to the reversal of the FDA's decision in 1983. The FDA was not about to be publicly embarrassed again.

Boehringer Ingelheim had a history of good relations with the AAAI, despite the 1983 FDA ruling. The company was a major sponsor of two nationally oriented educational programs, one of which, Asthma Care Training, was run by the AAAI and affiliated organizations. So when Boehringer Ingelheim wrote to the AAAI and asked for its help, its executive committee members agreed. Ironically the AAAI had just issued a position statement recommending that beta agonist inhalers continue to be sold on a prescription basis only.[83] At a meeting in Boston on 8 July 1985, six AAAI members, two Boehringer Ingelheim executives, and two Boehringer Ingelheim consultants gathered to review evidence on the OTC safety of Alupent. The fact that four out of ten members of the AAAI's ad hoc committee had ties to Boehringer Ingelheim, which was invested heavily in the meeting's outcome, raised serious ethical questions about conflict of interest.

It is not clear who took the minutes of the meeting, but Boehringer Ingelheim had to be pleased by its outcome. According to the minutes, the "entire Ad Hoc Committee" voted and "unanimously accepted [that] Alupent MDI is safe for OTC availability for adult use." Furthermore, the committee suggested that "Academy support of Alupent . . . will help maintain high product standards after the Boehringer patent expires in January, 1986."[84]

In the months to follow, all hell broke loose over what had really taken place at that meeting. One committee member and AAAI representative, Paul Hanaway, wrote to the general manager of Boehringer Ingelheim to complain that "in no way did [he] offer a vote of confidence for the safety of OTC Alupent MDI." The question posed to committee members, Hanaway recalled, was whether Alupent was safe. "No formal vote," he insisted, was taken to "approve or disapprove of OTC availability."[85] Shortly after the July meeting, William Pierson, AAAI co-chair of the ad hoc committee, told *Family Practice News,* "It is my opinion that these beta agonists (metaproterenol) should be prescribed by a physician rather than be available over the counter."[86] Boehringer Ingelheim immediately wrote to the other ad hoc committee co-chair and AAAI representative, Albert Sheffer, expressing concern and alarm.[87] In the weeks to follow, Boehringer Ingelheim representatives played a major role in shaping the summary statement approved by the ad hoc committee members on 12 August 1985. "MDI Metaproterenol was safe and effective," the statement concluded, "and as such was appropriate for the OTC market." But in an abrupt, unexplained turn of events, the final ad hoc committee report included no statement explicitly approving Alupent for OTC use.[88] In November 1985, the executive committee of the AAAI voted unanimously, despite the ad hoc committee's earlier recommendation, to reaffirm its original position statement, which had come out against OTC sales of beta agonist inhalers.

The OTC bronchodilator issue had become a political nightmare for the AAAI. Some prominent members of the academy spoke out in favor of the OTC distribution of Alupent. Others lobbied heavily against it. One thing was certain. As John Anderson, secretary of the AAAI, astutely observed in a thoughtful letter to AAAI president John Salvaggio, the field of clinical allergy had entered a whole new relationship to the pharmaceutical industry. "We have become very dependent upon industry," Anderson wrote, "for not only support of social events at our annual meeting (which we have discussed so often) but, also, for educational support—both for our society and for the Asthma and Allergy Foundation, which we sponsor. We are now entering into a phase where our marketing efforts will at least be partially dependent upon industry support. In addition, much of membership is reliant upon industry to support their clinical research programs, to partially support their private practices and to supplement their personal income—either through honoraria or trips." What was to be the "relationship of the AAAI as a society and industry in general?" Anderson asked. And what was to be "the relationship of individual members of the AAAI as they act on behalf of the AAAI and industry?" "I hope," Anderson concluded, "that no industry ever holds the AAAI hostage on any major policy decision."[89] Despite Anderson's hope, industry would try in subtle and not so subtle ways to sway the AAAI in directions that could greatly influence the share prices of particular drug companies. This was particularly evident in the aftermath of the OTC bronchodilator controversy.

By the mid-1980s, disturbing data were coming in that suggested asthma mortality rates in the United States were on the rise, despite a good number of "new and improved drugs." Death rates from asthma had jumped from 0.8 per 100,000 people in 1977 to 1.7 per 100,000 in 1985. Among African Americans, the rate had increased from 1.9 to 2.5

per 100,000 between 1979 and 1982. The rising death toll was taking place against the background of a marked increase in the sales of anti-asthma drugs, particularly beta agonist inhalers. Whatever the cause of the apparent increase in death rates, the data suggested caution with respect to bronchodilator use. At the same November 1985 meeting in which the executive committee of the AAAI reaffirmed its position against the OTC sale of beta agonist inhalers, the committee recommended the appointment of a task force on asthma mortality. The task force comprised representatives from governmental agencies, pharmaceutical companies, and the medical profession who were charged with assessing the current causes and conditions of asthma deaths.[90]

When the AAAI approached Boehringer Ingelheim for help in funding the Task Force on Asthma Mortality, the drug company was less than enthusiastic. Repeated studies over two decades, it insisted, had made it "clear that asthma mortality is not related to the use of beta-agonists." Patient behavior and education, it argued, not drugs, were the problem.[91] Meanwhile, GlaxoSmithKline enthusiastically wrote a $10,000 check in support. GlaxoSmithKline had lobbied strongly against FDA approval of OTC sales of beta agonist inhalers. OTC status, it feared, would place "the treatment of asthma, and possibly its diagnosis, in the hands of the layperson, which could result in potential life-threatening situations."[92] But self-medication would also cut significantly into the profits of its own prescription asthma drugs. Schering Corporation, whose patent on Proventil would not expire until 1989, also wrote a generous check in support of the Task Force on Asthma Mortality.[93] Only after AAAI member Albert Sheffer assured Boehringer Ingelheim that metaproterenol was "not the issue" in asthma mortality, "but developing a program of education" was, did the company agree to support the Task Force.[94] As the stake in asthma drug profits grew, so too did the extent to which drug companies played hardball in an effort to win consumers and physicians.

An escalating presence of drugs in allergic bodies increased profits to pharmaceutical companies in the 1980s, but the drugs did little to stem the rising tide of asthma. Self-reported prevalence rates had increased 29 percent between 1980 and 1987. Hospitalization rates for children with asthma increased 4.5 percent annually during the 1980s.[95] This trend was all the more disturbing since total hospitalization rates for children were on the decline. How much increased marketing and public awareness accounted for the sharp rise in prevalence was anybody's guess. But the disconcerting trend in asthma morbidity and mortality prompted the medical community to reassess the direction of drug treatment and therapy for asthma over the previous decade.

In 1989, the National Heart, Lung, and Blood Institute's National Asthma Education Program (NAEP) convened an expert panel to develop a series of guidelines for the diagnosis and management of asthma in light of the recent statistical trends. Medical experts in the field of allergy and asthma, much as they had done in response to the bronchodilator scare of the 1960s, emphasized overreliance on a single drug—namely, bronchodilators—as a contributing factor in the rise of asthma hospitalization and mortality rates in the 1980s. The experts further concluded that asthma was underdiagnosed and often inadequately treated. They emphasized a more integrated, comprehensive strategy to asthma management and care that brought together environmental control measures, drug treatment, and patient education.[96]

It was the new drug therapies coming out of the NAEP guidelines that attracted the most media attention.[97] According to the NAEP expert panel, far too little attention had been paid to the role of chronic airway inflammation in the pathology of asthma. Bronchodilator therapy had underscored the definition of asthma as reversible airway obstruction. In the early 1990s, new drugs became linked to asthma's reclassification as a chronic inflammatory disease of the airways. Suddenly, a class of anti-inflammatory drugs—inhaled steroids—moved to center stage as the next wonder drugs in the treatment of asthma.[98]

Such a change in disease definition and treatment could be a huge economic boon to particular pharmaceutical firms. For example, in the early 1980s, Forest Laboratories was a small generic company. Its main allergy and asthma drug was Theochron, a generic slow-release theophylline. Theochron faced a significant uphill battle to gain market shares from Schering's product, Theo-Dur, which outsold Theochron three to one. Schering was a giant in the respiratory allergy field. When Schering acquired Key Pharmaceuticals and its drug Theo-Dur in 1986, it put considerable pressure on the AAAI's Committee on Drugs to issue a policy statement against pharmacist substitution of one slow-release theophylline drug for another without a doctor's permission. A "Do Not

Substitute" policy statement would have been a ringing endorsement of Theo-Dur. Forest Laboratories heavily lobbied the AAAI to come out against a "Do Not Substitute" recommendation. In the end, the executive committee of the AAAI decided in 1988 not to publish any policy statement regarding product substitution of slow-release theophylline. Members of the AAAI's Committee on Drugs who supported a "Do Not Substitute" position statement expressed dismay at how a medically oriented issue had become "an issue of economics and marketing."[99] Meanwhile, Forest Laboratories, grateful that the AAAI had refused to take an official position on theophylline, inquired about how it might become a greater sponsor of academy activities.[100]

In an era when clinical allergy had become enmeshed in the world of billion-dollar drugs, no medical issue was free of economics and marketing. Any policy statement, or even the lack of one, could be taken by drug companies as a product endorsement. This was certainly how Forest Laboratories viewed NAEP guidelines published in February 1991 that underscored asthma as an inflammatory disease. In its 1991 annual report, the company announced to stockholders in bold headlines, "Forest's Respiratory Products Recommended in New NIH Asthma Treatment Guidelines." Thanks to the recommendation of the NAEP expert panel, which promoted greater clinical use of inhaled steroids in asthma treatment, prescriptions of Aerobid, Forest Laboratories' inhaled steroid, soared. In the first three months of 1991, prescriptions of Aerobid doubled those of the entire previous year. Aerobid was the fastest growing inhaled steroid on the market. Aerochamber, also made by Forest, became the best-selling spacer device for delivering inhaled steroids in the United States. By 1991, Forest Laboratories had a total market value of $1.5 billion, an impressive gain from its value a decade before of little more than $10 million.[101] Other drug companies rushed to cash in on the latest wave of allergy/asthma medications that would promise allergy sufferers escape from the limitations of their illness. Better drugs. Better profits. Better lives? Mortality and morbidity rates of asthma in the United States continued to climb.

Each morning, when my son and I breathe in fluticasone proprionate, GlaxoSmithKline's inhaled steroid, we line our bronchial airways with a chemical intended to insulate us from the irritating effects of allergens that connect our bodies to the places in which we live. As consumers, we buy into the idea of escape from place. But unlike the hay feverites of the late 1800s, we no longer have to flee our location. Instead, we fortify our allergic bodies with drugs. The drugs let us get on with our lives, to perform at work and play without regard to our environment. It is an alluring solution. How much easier to take a pill or a puff than to move to the mountains or lakeshore or desert—so much easier than addressing issues of land use, rethinking building construction, or confronting structural inequities in housing and health care in American society. We take a pill or a puff, feel better, and conveniently ignore how that chemical moving inside our bodies connects us to a larger political economy and ecology of allergic disease.

In this economy, drugs and diseases are intimately related, medicine and commerce are inseparable, and consumption and performance are inextricably linked. In this economy, the flow of drugs benefits some more than others. Minority children, most severely affected by asthma, were the last to benefit from the introduction of inhaled steroids.[102] For underprivileged urban sufferers of asthma, the cough and cold section of the local store or the hospital emergency room were far more likely to be the places of treatment and dispensers of drugs than the clinical allergist's office. Xolair, Novartis's newly touted anti-IgE wonder drug for asthma, at an estimated cost of $12,000 per year, isn't likely to find its way into uninsured, asthma-stricken, inner-city neighborhoods any time soon. As the overuse of bronchodilators made clear, the drugs delivered to our lungs can have negative ecological effects on our bodies. The chemicals used to deliver those drugs have also affected the ecology of the planet. Until recently, bronchodilator inhalers and inhaled steroid dispensers not only delivered drugs to counter asthma symptoms but they also expelled chlorofluorocarbons. The Montreal Protocol initiated the phasing out of these chemicals in 1989 because of their destructive effects on the ozone layer in the Earth's upper atmosphere.

We have too narrowly focused our attention on the body within. We have done so because we have been lulled into complacency by symptomatic relief. But also we have been propelled by the momentum of corporate capital and biomedical research, a powerful combination that continuously holds out the euphoric promise of the next magic bullet, a cure for all our ills. But there is no magic cure-all, only a price to pay.

That price is the failure to recognize and benefit from understanding the evolving relationship of body and place—that is, the body as part of, not isolated from, an environment to which it is continually responding. In looking back on the path we've taken, we can illuminate the changing relationships, the interplay of disease, environment, and place that has shaped the history of allergy in modern America.

Epilogue

In 1969, at the twenty-fifth anniversary meeting of the American Academy of Allergy, René Dubos stepped up to the platform to give the Robert A. Cooke Memorial Lecture. The speaker, who would become best known for coining the phrase "Think Globally, Act Locally," seemed an odd choice to deliver the academy's most esteemed lecture. Dubos was neither a trained physician nor a leading immunologist. In fact, never in his professional career as a microbiologist at Rockefeller University had Dubos published an article on the subject of allergy. Heralded for his isolation of germ-fighting drugs produced by soil microbes—a discovery that ushered in the modern era of antibiotics—Dubos became quite skeptical in the 1950s of what others saw as the golden age of medicine. Physicians had adopted a "naïve cowboy philosophy," Dubos wrote in his 1959 book, *The Mirage of Health;* it was evidenced in their belief that drugs spelled the conquest of disease. Such a belief was founded on the doctrine "that each disease had a well-defined cause and that its control could best be achieved by attacking the causative agent." But the doctrine failed to "take account of the difficulties arising from the ecological complexity of human problems."[1] Ten years later, Dubos told the crowd of allergists gathered at the American Academy of Allergy meeting, "Physicians and biologists study in great detail bodily structures and functions. Technologists develop ingenious and powerful techniques for manipulating nature and creating artifacts. But hardly any attention is paid to the interplay between man and the world in which he lives."[2]

Dubos looked upon disease not as some external agent, a demon that invaded the human body. Rather, he, like his contemporary, the

Australian virologist MacFarlane Burnet, regarded "disease as a manifestation of the interaction of living beings."[3] Disease was to be seen as a symptom, not a cause, of changing relationships between organisms and their environments. Medicine, Dubos urged the allergists in 1969, needed once again to turn its attention to the organism in relation to its surroundings and way of life. That, he reminded them, was, after all, the idea behind allergy's etymological meaning: altered reactivity. "All aspects of life," Dubos argued, "can be considered as manifestations of allergy," because all life modified itself in relation to its changing surroundings.[4]

Although this ecological vision of disease gained little traction in an age of miracle cures and wonder drugs, the return of malaria, cholera, and tuberculosis and the threat of emerging diseases such as AIDS, the Ebola virus, and avian influenza have brought renewed interest in ecological perspectives on disease, particularly among population health researchers. What ecology means in this context varies widely. Harvard biologist Richard Levins articulates one ecological approach, which he has called an ecosocial view. Levins's view is built on the "inseparability of social, ecological, physical, chemical and biotic environments." "Health," he and his colleague Cynthia Lopez write, "is produced and eroded in a natural and social environment that varies in time and space and according to the social locations of people in various hierarchical, cooperative, and competitive relationships." Disease is a product of complex biological, economic, material, and social relations shifting in both space and time.[5]

Allergy is not a thing but a relation. It is a way of being in the world that changes in both place and time. Each place we encounter is the historical product of a certain set of material, economic, and social relations. Those relations give form and substance to the experience, meaning, and understanding of allergic disease. Economic and social forces that helped turn nature into leisure in the White Mountains or along the shores of Lake Michigan also helped to define hay fever as a luxurious disease of wealthy urbanites. The making of urban wastelands and fallow fields transformed ragweed into public health enemy number one and forged new relationships between physicians and plant ecologists in the production of knowledge about allergic disease. The building of Tucson as a health capital of the world turned the dry desert air into a resource sought after by asthmatics struggling to breathe, people who saw the desert as their last hope and fought in vain to preserve its therapeutic qualities. Economic, environmental, and social inequities of the urban ghetto made asthma a far more serious disease than in white middle-class suburbs of America and turned asthma into a focal point of political action and protest. Changing relationships of buildings materials, engineering design, home décor, and indoor behavior patterns transformed the image of the home first into a safe haven free from allergy and then into an ecological nightmare. The tightening relationships among bodies, drugs, markets, medicine, consumption, and performance in the postwar American economy made the chemical alteration of the immune landscape part of the everyday experience of most contemporary allergy sufferers.

Taken as a whole, the stories of place and illness told in this book make clear that allergy cannot be isolated as a disease. It is not separate from the complex of environmental relations—physical, social, economic—out of which it came into being. This is precisely the danger of "equating disease with the effect of a precise cause," about which Dubos warned.[6] But this is the path that modern medicine took. In doing so, medicine negated a relational concept of health that emphasized not things in themselves—germs or allergens, for example—but evolving interactions among people, other living organisms, and inanimate matter and energy. History and geography, informed by an ecological approach to disease, make visible how those interactions came together in the making of allergic illness. They also make visible the variability of those relationships, as allergy in different places and different times has come to have an impact on human lives in quite different, and often unjust, ways. Place, the historical product of relationship, becomes a focal point, a nexus where the interrelationships between bodies and environments are constantly being made and remade.

Relationship matters. Ecology, in that it is the study of relationship, matters. Place matters: in the making, in the experience, and in the knowledge of illness and health.

NOTES

MANUSCRIPT COLLECTIONS

AAAAAI	Archives of the American Academy of Allergy, Asthma, and Immunology, University of Wisconsin, Milwaukee
AIHP	American Institute of the History of Pharmacy, University of Wisconsin, Madison
BVA	Bay View Association Archives, Bay View, Michigan
GASP	Group against Smelter Pollution, University of Arizona Library, Special Collections, Tucson, Arizona
HMHP	Harvey Monroe Hall Papers, Bancroft Library, University of California, Berkeley
JMP	Jeremiah Metzger Papers, 1906–1953, MS 490, Special Collections, University of Arizona Library
LNGP	Leslie N. Gay Papers, Alan Chesney Medical Archives, Johns Hopkins University
NAC	National Asthma Center Archives, Special Collections, University of Denver
OSUA	Oregon State University Archives, Oregon State University
RBSC	Rare Books and Special Collections, Francis A. Countway Library of Medicine, Harvard University
SC-NYPL	Schomburg Collection, New York Public Library
UASC	Main Library, Special Collections, University of Arizona

PREFACE

1. American Academy of Allergy, Asthma, and Immunology, *Overview of Allergic Diseases,* vol. 1 of *The Allergy Report* (Milwaukee: American Academy of Allergy, Asthma, and Immunology, 2000), i.

2. Warren T. Vaughan, *Strange Malady: The Story of Allergy* (New York: Doubleday, 1941).

INTRODUCTION

1. "Mr. T Chops Away at Lake Forest's Fiber," *Chicago Tribune,* 22 May 1987, 1, 12; "Genteel Chicago Suburb Rages over Mr. T's Tree Massacre," *New York Times,* 30 May 1987, 8.

2. "Protecting Trees in Lake Forest," *New York Times,* 25 October 1987, 49.

3. "Mr. T Chops Down Trees on His Property; Angers His Neighbors in Suburb," *Jet* 72 (8 June 1987): 57.

4. Mr. T, *Mr. T: The Man with the Gold* (New York: St. Martin's Press, 1984).

5. Sandra D. Thomas and Steven Whitman, "Asthma Hospitalizations and Mortality in Chicago," *Chest* 116 (1999): 135S–141S.

6. For a history of Lake Forest, see Kim Coventry, Daniel Meyer, and Arthur H. Miller, *Classic Country Estates of Lake Forest: Architecture and Landscape Design, 1856–1940* (New York: W. W. Norton, 2003).

7. William Cronon's *Nature's Metropolis: Chicago and the Great West* (New York: W. W. Norton, 1991) offers a brilliant history of the transformation of urban and rural landscape in the making of Chicago.

8. George M. Beard, *Hay-Fever; or, Summer Catarrh: Its Nature and Treatment* (New York: Harper, 1876), 87.

9. John Muir, *Our National Parks* (Boston: Houghton, Mifflin, 1901), 1. Quoted in Charles W. Eliot, *Charles Eliot: Landscape Architect* (1902; repr. Amherst: University of Massachusetts Press, 1999), 341.

10. Frederick Law Olmsted, *Public Parks and the Enlargement of Towns* (1870; New York: Arno Press, 1970), 17.

11. *Report of the Special Park Commission to the City Council of Chicago on the Subject of a Metropolitan Park System* (W. J. Hartman, 1905), 133; Olmsted, *Public Parks and the Enlargement of Towns,* 15.

12. Robert E. Grese, *Jens Jensen: Maker of Natural Parks and Gardens* (Baltimore: Johns Hopkins University Press, 1992), 26, 9.

13. "Halts Hay Fever by Conditioned Air," *New York Times,* 7 May 1933, 31.

14. American Lung Association, Epidemiology and Statistics Unit, Research and Program Services, *Trends in Asthma Morbidity and Mortality,* May 2005; Pew Environmental Health Commission, *Attack Asthma: Why America Needs a Public Health Defense System to Battle Environmental Threats* (2000); available at http://www.pewtrusts.com/pubs/pubs_item.cfm?content_item_id=310&content_type_id=8&page=p1.

CHAPTER 1: HAY FEVER HOLIDAY

1. Quoted in Morrill Wyman, *Autumnal Catarrh (Hay Fever)* (New York: Hurd and Houghton, 1872), 57.

2. Ibid., 49. See also "Outing in Hay Fever," *White Mountain Echo,* 31 August 1889, 5. On the body as an instrument of environmental measurement,

see Michael Dettelbach, "The Face of Nature: Precise Measurement, Mapping, and Sensibility in the Work of Alexander von Humboldt," *Studies in the History and Philosophy of Biological and Biomedical Sciences* 30 (1999): 473–504.

3. Paxton Hibben, *Henry Ward Beecher: An American Portrait* (New York: Press of the Readers Club, 1942), vii.

4. Quoted in Wyman, *Autumnal Catarrh,* 143, 55.

5. "Hay Fever Day," *White Mountain Echo,* 7 September 1878, 3; Harrison Rhodes, "American Holidays: Springs and Mountains," *Harper's Monthly* 129 (1914): 545; George M. Beard, *Hay-Fever; or, Summer Catarrh: Its Nature and Treatment* (New York: Harper, 1876), 82.

6. George M. Beard, *American Nervousness: Its Causes and Consequences* (New York: G. P. Putnam's Sons, 1881), 7. On Beard and neurasthenia, see Charles Rosenberg, "Pathologies of Progress: The Idea of Civilization as Risk," *Bulletin of the History of Medicine* 72 (1998): 714–730; George Frederick Drinka, *The Birth of Neurosis: Myth, Malady and the Victorians* (New York: Simon and Schuster, 1984); F. G. Gosling, *Before Freud: Neurasthenia and the American Medical Community, 1870–1910* (Chicago: University of Illinois Press, 1987); Barbara Sicherman, "The Uses of a Diagnosis: Doctors, Patients, and Neurasthenia," *Journal of the History of Medicine and Allied Sciences* 32 (1977): 33–54.

7. J. Bostock, "Of Catarrhus Aestivus," *Medico-Chirurgical Transactions* 14 (1828): 437–446. On the history of hay fever, see M. B. Emanuel, "Hay Fever, a Post Industrial Revolution Epidemic: A History of Its Growth during the 19th Century," *Clinical Allergy* 18 (1988): 295–304; Kathryn J. Waite, "Blackley and the Development of Hay Fever as a Disease of Civilization in the Nineteenth Century," *Medical History* 39 (1995): 186–196.

8. Charles H. Blackley, *Experimental Researches on the Nature and Causes of Catarrhus Aestivus (Hay-Fever, or Hay-Asthma)* (London: Baillière, Tindall, and Cox, 1873), 7.

9. Beard, *Hay-Fever,* 86, 81.

10. Blackley, *Experimental Researches,* 155.

11. Ibid., 162.

12. Morrell Mackenzie, *Hay Fever and Paroxysmal Sneezing: Their Etiology and Treatment,* 4th ed. (London, 1887), 10.

13. William Hard, "The Great American Sneeze," *Everybody's Magazine* 25 (1911): 262–263.

14. Webster discussed and quoted in Wyman, *Autumnal Catarrh,* 138.

15. Ibid., 137.

16. Samuel P. Lyman, *Life and Memorials of Daniel Webster,* vol. 2 (New York: D. Appleton, 1853), 63. The most comprehensive biography of Webster that includes discussion of his illness is Robert V. Remini, *Daniel Webster: The Man and His Time* (New York: W. W. Norton, 1997).

17. Webster discussed and quoted in Wyman, *Autumnal Catarrh,* 137–139.

18. John F. Sears, *Sacred Places: American Tourist Attractions in the Nineteenth Century* (New York: Oxford University Press, 1989), 49. See also Dona Brown, *Inventing New England: Regional Tourism in the Nineteenth Century* (Washington, D.C.: Smithsonian Institution Press, 1995).

19. On the religious symbolism of the Willey disaster, see Sears, *Sacred Places,* 72–86; Eric Purchase, *Out of Nowhere: Disaster and Tourism in the White Mountains* (Baltimore: Johns Hopkins University Press, 1999), 4–21.

20. "Hay Fever," *White Mountain Echo,* 30 August 1879, 3.

21. Helen Hunt, "Mountain Life: The New Hampshire Town of Bethlehem," *New York Evening Post,* 18 October 1865, 1.

22. For estimates on the number of hay fever residents, see "A Warning to Bethlehem," *White Mountain Echo,* 14 September 1889, 7; Thomas F. Anderson, "Our New England as a National Health Resort," *New England Magazine* 38 (1908): 313–327.

23. Morrill Wyman, "Journey to the White Mountains to Avoid Autumnal Catarrh" (1867), B MS b 200.1, RBSC.

24. "The Beauties of Hay Fever," *White Mountain Echo,* 19 September 1891, 2.

25. Quoted in "Hay Fever," *White Mountain Echo,* 5 September 1891, 8.

26. "Is Hay Fever a Fiction?" *White Mountain Echo,* 10 September 1881, 8. On gout as an upper-class disease, see Roy Porter and G. S. Rousseau, *Gout: The Patrician Malady* (New Haven: Yale University Press, 1998).

27. Quoted in "Hay Fever," *White Mountain Echo,* 30 August 1879, 3.

28. On the nineteenth-century class aesthetic of consumption, see Kathlerine Ott, *Fevered Lives: Tuberculosis in American Culture since 1870* (Cambridge: Harvard University Press, 1996), 13–16.

29. Quoted in "Hay Fever," *White Mountain Echo,* 1 September 1888, 5.

30. For the popularity of these remedies, see, e.g., *White Mountain Echo,* 12 September 1885, 5, and 6 September 1890, 13; *Manual of the United States Hay-Fever Association for 1887* (Lowell, Mass., 1887).

31. E. J. Marsh, "Hay Fever, or Pollen Poisoning," *Transactions of the Medical Society of New Jersey, Newark,* 1877, 86–109. In the medical literature Wyman, *Autumnal Catarrh,* 82, and John O. Roe, "Coryza Vasomotoria Periodice (Hay Asthma) in the Negro," *Medical Record* 26 (1884): 427–428, report the only two cases of hay fever among blacks and Native Americans that I have found in late nineteenth-century America.

32. "Broken Planks Again," *White Mountain Echo,* 28 August 1880, 2.

33. *Daily Resorter,* 3 September 1894, 6.

34. "Hay Feverites Provoked," *White Mountain Echo,* 14 September 1889, 8.

35. On hotel proprietors' concerns, see "The Cure of Hay Fever," *White Mountain Echo,* 30 August 1884, 9.

36. Rate comparisons to New York City hotels are from Purchase, *Out of Nowhere,* 52. Rates for the Maplewood Hotel and Sinclair House are from "White Mountain Resorts," *White Mountain Echo,* 7 August 1880, 4.

37. Karl P. Abbott, *Open for the Season* (New York: Doubleday, 1950), 19.

38. *Profile House, Franconia Notch, White Mountains, N.H.: A Summer's Sojourn* (New York, 1895), 13–14.

39. See, e.g., Leo Marx, *The Machine in the Garden: Technology and the Pastoral Ideal in America* (New York: Oxford University Press, 1964); Ann Whiston Spirn, "Constructing Nature: The Legacy of Frederick Law Olmsted," in *Uncommon Ground: Toward Reinventing Nature,* ed. William Cronon (New York: W. W. Norton, 1995), 91–113; Richard White, *The Organic Machine: The Remaking of the Columbia River* (New York: Hill and Wang, 1995).

40. "The White Mountains: Their Popularity Increased by the Facilities Afforded for Viewing Them," *White Mountain Echo,* 2 July 1878, 1.

41. Peter J. Schmitt, *Back to Nature: The Arcadian Myth in Urban America* (New York: Oxford University Press, 1969), 5.

42. On the impact of the railroads on grain markets and regional economics, see Purchase, *Out of Nowhere,* 49. The creation of the White Mountains as a tourist region for major metropolitan centers of the northeast parallels the story of the creation of Chicago's hinterlands by the railroads in William Cronon, *Nature's Metropolis: Chicago and the Great West* (New York: W. W. Norton, 1991). Far too little attention has been paid to date to the integral role that railroads played in the development of regional health economies in nineteenth-century America.

43. "Is Hay Fever a Fiction?" *White Mountain Echo,* 10 September 1881, 8.

44. Cronon, *Nature's Metropolis.*

45. Robert W. Rydell, *All the World's a Fair: Visions of Empire at American International Expositions* (Chicago: University of Chicago Press, 1984), 39–40, 45.

46. Quoted in Jeffrey Meyers, *Hemingway: A Biography* (New York: Harper and Row, 1985), 4.

47. Estimate based on reports in the *Daily Resorter.* See, e.g., "Still Another Thousand," *Daily Resorter,* 30 August 1895.

48. Nancy T. Harvis, ed., *Historical Glimpses: Petoskey* (Petoskey, Mich.: Little Traverse Historical Society, 1986).

49. On Michigan's timber industry and forest conservation, see Barbara E. Benson, *Logs and Lumber: The Development of the Lumber Industry in Michigan's Lower Peninsula, 1837–1870* (Mount Pleasant: Central Michigan University, 1989); Dave Dempsey, *Ruin and Recovery: Michigan's Rise as a Conservation Leader* (Ann Arbor: University of Michigan Press, 2001).

50. *Souvenir of Petoskey Homecoming: A History of the Past and Present* (Petoskey, Mich.: Petoskey Booster Club, 1938).

51. "The Final Meeting," *Daily Resorter,* 14 September 1892.

52. "Caught on the Fly," *Daily Resorter,* 2 September 1893.

53. "The Final Meeting"; "Hay Fever Convention," *Emmet County Democrat,* 8 September 1882; "Convention Tomorrow," *Daily Mining Journal,* 29 August 1899.

54. "Constitution and By Laws," *Daily Resorter,* 11 August 1890; "Ah-h Kerchoo!!" *Daily Resorter,* 29 August 1890.

55. "Not to Be Sneezed At," *Daily Resorter,* 28 August 1890.

56. "Hay Fever Convention," *Emmet County Democrat,* 16 September 1881.

57. *Daily Resorter,* 4 September 1894, 8.

58. *Daily Resorter,* 19 August 1893, 3.

59. *Summer Tours on Great Lakes* (Chicago: Northern Michigan Transportation, 1901).

60. See John F. Reiger, *American Sportsmen and the Origins of Conservation* (New York: Winchester Press, 1975).

61. Arthur Holbrook, "Hay Fever Vacations; or Jaunts and Camps in Northern Wisconsin for Health and Recreation"; reprinted from the *American Angler* (Milwaukee: Swann and Tate, 1886), 8, 22–23.

62. "Auditorium Acceptance 1915 Folder," p. 5; Individuals—Smith, Clement; BVA.

63. Quoted in Keith J. Fennimore, *The Heritage of Bay View, 1875–1975* (Grand Rapids, Mich.: William B. Eerdmans, 1975), 21.

64. On the history of Bay View, see ibid. and John A. Weeks, *Beneath the Beeches: The Story of Bay View, Michigan* (Grand Rapids, Mich.: William B. Eerdmans, 2000).

65. "Hay Fever Convention," *Emmet County Democrat,* 8 September 1882.

66. "Hay Fever Meeting," *Daily Resorter,* 25 August 1896. The election of a woman president was not without controversy but was supported by John M. Hall, assembly director at Bay View Association. See "Annual Meeting," *Daily Resorter,* 6 September 1895.

67. "The Newport of Chicago Society," *The 400* 2 (31 May 1894).

68. Quoted in Mackinac State Historic Parks, *Mackinac: An Island Famous in These Regions* (Mackinac Island, Mich.: Mackinac State Historic Parks, 1998), 51.

69. J. A. Van Fleet, *Summer Resorts of the Mackinaw Region, and Adjacent Localities* (Detroit: Lever Press, 1882), 18.

70. On Saratoga and Newport, see John Sterngrass, *First Resorts: Pursuing Pleasure at Saratoga Springs, Newport and Coney Island* (Baltimore: Johns Hopkins University Press, 2001).

71. For a history of the Grand, see John McCabe, *Grand Hotel: The Biography of an American Institution* (Mackinac Island, Mich.: Unicorn Press, 1987). See also Phil Porter, "1890s: Mackinac Island's Decade of Change," *Michigan History Magazine,* July–August 1995, 9–17.

72. "The Newport of Chicago Society"; "Respite for Hay Fever Patients," *Chicago Evening Post,* 15 August 1896.

73. Grand Rapids and Indiana Railroad, *The Summer Resorts and Waters of Northern Michigan* (Chicago: Poole Bros., 1885), 6, 25. On the romanticiza-

tion of the American Indian during this period, see Shepard Krech III, *The Ecological Indian: Myth and History* (New York: W. W. Norton, 1999).

74. Ernest Hemingway, "A Big Two-Hearted River," in *The Complete Short Stories of Ernest Hemingway,* Finca Vigia Edition (New York: Charles Scribner's Sons, 1987), 164. See also Fredrick Chr. Brogger, "Whose Nature? Differing Narrative Perspectives in Hemingway's 'Big Two-Hearted River,' " in *Hemingway and the Natural World,* ed. Robert E. Fleming (Moscow: University of Idaho Press, 1999), 19–30.

75. "Bethlehem's Progress," *White Mountain Echo,* 14 September 1878, 1. On the influence of hay fever sufferers on the tourist season, see "Fun about Hay Fever," *White Mountain Echo,* 28 August 1880, 2. On tourists as outsiders, see John A. Jakle, *The Tourist: Travel in Twentieth-Century America* (Lincoln: University of Nebraska Press, 1985).

76. Wyman, *Autumnal Catarrh,* 95.

77. Ibid., 127.

78. "Montpellier, Vt.," *White Mountain Echo,* 5 August 1882, 5.

79. See, e.g., "Bethlehem at Work," *White Mountain Echo,* 17 September 1881, 8; "Hay Fever," *White Mountain Echo,* 10 September 1881, 8; "Hay Fever," *White Mountain Echo,* 11 September 1880, 3.

80. "Bethlehem Association," *White Mountain Echo,* 11 September 1880, 7. See also "Down with the Dust," *White Mountain Echo,* 3 August 1878, 3.

81. "With Weeping Eyes," *Daily Resorter,* 17 September 1890, 4.

82. "Hay Fever," *White Mountain Echo,* 3 September 1881, 5; *White Mountain Echo,* 10 September 1881, 2; "Hay Fever," *White Mountain Echo,* 7 September 1895, 2. For a vivid description of how hay fever determined the daily path of one sufferer, see Beard, *Hay-Fever,* 199.

83. Wyman, *Autumnal Catarrh,* 101.

84. Quoted in "U.S. Hay-Fever Association," *White Mountain Echo,* 18 September 1886, 6; "Ragweed on the Railroad," *White Mountain Echo,* 14 September 1878, 3.

85. James Eugene Bell, "Prize Essay. II," *Manual of the United States Hay Fever Association for 1887,* 30.

86. Quoted in "Hay Fever Experiences," *White Mountain Echo,* 12 September 1896, 8.

87. "Hay Fever Talk," *White Mountain Echo,* 10 September 1892, 8.

88. "The United States Hay Fever Association," *White Mountain Echo,* 8 September 1900, 19.

89. Samuel Lockwood, "The Comparative Hygiene of the Atmosphere in Relation to Hay Fever," *Journal of the New York Microscopical Society* 5 (1889): 50.

90. On the importance of tuberculosis to the agrarian order, see Georgina D. Feldberg, *Disease and Class: Tuberculosis and the Shaping of Modern North American Society* (New Brunswick, N.J.: Rutgers University Press, 1995), 11–35, 53.

91. *The Hay Fever and Where to Find Relief* (Marquette, Mich.: Mining Journal, 1877), 5.

92. Henry David Thoreau, *The Writings of Henry David Thoreau,* ed. Bradford Torrey (Boston: Houghton, Mifflin, 1906), 11:9, 11. On the pastoral ideal in Thoreau's writings, see Lawrence Buell, *The Environmental Imagination: Thoreau, Nature Writing, and the Formation of American Culture* (Cambridge: Belknap Press, 1995). For a classic work on the important cultural and environmental significance of the country/city divide, see Raymond Williams, *The Country and the City* (New York: Oxford University Press, 1973). For a provocative analysis suggestive of the relationships between disease and Thoreau's bodily engagement with the environment, see Christopher Sellers, "Thoreau's Body: Towards an Embodied Environmental History," *Environmental History* 4 (1999): 486–514.

93. On wilderness as the therapeutic balm for late-nineteenth- and early-twentieth-century middle- and upper-class urban Americans, see William Cronon, "The Trouble with Wilderness; or Getting Back to the Wrong Nature," in Cronon, ed., *Uncommon Ground,* 69–90; T. Jackson Lears, *No Place of Grace: Antimodernism and the Transformation of American Culture, 1880–1920* (New York: Pantheon, 1981); Gregg Mitman, *Reel Nature: America's Romance with Wildlife on Film* (Cambridge: Harvard University Press, 1999); Schmitt, *Back to Nature.*

94. Quoted in "Hay Fever," *White Mountain Echo,* 13 September 1879, 1.

95. See, e.g., Kenneth Thompson: "Wilderness and Health in the Nineteenth Century," *Journal of Historical Geography* 2 (1976): 145–161, and "Trees as a Theme in Medical Geography and Public Health," *Bulletin of the New York Academy of Medicine* 54 (1978): 517–531.

96. Quoted in "Hay Fever Meeting," *White Mountain Echo,* 2 September 1899, 17.

97. *Manual of the United States Hay Fever Association for 1900* (1900), 39. On the history of forest conservation in the White Mountains, see Richard W. Judd, *Common Lands, Common People: The Origins of Conservation in Northern New England* (Cambridge: Harvard University Press, 1997), 91–112.

98. "Hay Fever Meetings," *White Mountain Echo,* 8 September 1894, 14.

99. *Manual of the United States Hay Fever Association for 1900,* 38. See also "Hay Fever Meetings," *White Mountain Echo,* 31 August 1889, 6; "A Round Dozen," *White Mountain Echo,* 21 September 1889, 7; "Flowers and Hay Fever," *White Mountain Echo,* 2 September 1893; "Forest Preservation," *White Mountain Echo,* 1 September 1894, 8; "Hay Fever Meeting," *White Mountain Echo,* 31 August 1895, 7; "Forestry," *White Mountain Echo,* 9 September 1899, 7; "Hay Fever Association," *White Mountain Echo,* 16 September 1899, 4.

100. Peter Rowan and June Hammond Rowan, *Mountain Summers* (Gorham, N.H.: Gulfside Press, 1995). Such physical freedom experienced by women hay fever sufferers in mountain resorts was similar to the sense of liberation expressed earlier by women seeking the water cure. See, e.g.,

Susan E. Cayleff, *Wash and Be Healed: The Water-Cure Movement and Women's Health* (Philadelphia: Temple University Press, 1987); Joan D. Hedrick, *Harriet Beecher Stowe: A Life* (New York: Oxford University Press, 1994), 173–185; Kathryn Kish Sklar, "All Hail to Pure Cold Water!" in *Women and Health in America: Historical Readings,* ed. Judith Leavitt (Madison: University of Wisconsin Press, 1984), 246–254.

101. *Report of the Forestry Commission of New Hampshire, January Session, 1891* (Manchester, 1891), 20.

102. Quoted in Judd, *Common Lands, Common People,* 104.

103. "Hay Fever Experiences," *White Mountain Echo,* 12 September 1896, 8; "Not to Be Sneezed At," *Denver Rocky Mountain News,* 25 August 1896. On Denison and medical climatology, see Billy M. Jones, *Health-seekers in the Southwest, 1817–1900* (Norman: University of Oklahoma Press, 1976), and Sheila M. Rothman, *Living in the Shadow of Death: Tuberculosis and the Social Experience of Illness in American History* (New York: Basic Books, 1994), 148–160.

104. "U.S. Hay Fever Association," *White Mountain Echo,* 17 September 1887; "Hay Feverites Provoked," *White Mountain Echo,* 14 September 1889, 8.

105. Lockwood, "The Comparative Hygiene of the Atmosphere," 49–55.

106. "The Worth of a Forest Patch," *White Mountain Echo,* 9 July 1892, 5.

107. "Hay Fever Meeting," *White Mountain Echo,* 2 September 1899, 17; *Manual of the U.S. Hay Fever Association for 1888* (Philadelphia, 1888), 18–19.

108. Hemingway, *The Complete Short Stories,* 516.

109. Dempsey, *Ruin and Recovery,* 56.

110. Horace Sutton, "The Scourge of the Sneeze," *Saturday Review of Literature,* 14 July 1951, 41–44; Frederick J. Vintinner and George W. Morrill, *Hay Fever Studies in New Hampshire 1947* (Concord, 1948).

111. "They Come to Sneeze and Remain to Play," *Mackinac Island News,* 21 August 1937, 1.

112. Dempsey, *Ruin and Recovery.* See also James Kates, *Planning a Wilderness: Regenerating the Great Lakes Cutover Region* (Minneapolis: University of Minnesota Press, 2001).

113. "Sneezers Flee," *Business Week,* 12 August 1944, 26.

114. "They Come to Sneeze and Remain to Play," 1.

CHAPTER 2: WHEN POLLEN BECAME POISON

1. C. von Pirquet and B. Schick, *Das Serumkrankheit* (Leipzig: Deuticke, 1905). See also Arthur M. Silverstein, "Clemens Freiherr von Pirquet: Explaining Immune Complex Disease in 1906," *Nature Immunology* 1 (2000): 453–455; Mark Jackson, *Allergy: The History of a Modern Malady* (London: Reaktion Books, 2006), 27–55.

2. Kenton Kroker, "Immunity and Its Other: The Anaphylactic Selves of Charles Richet," *Studies in the History and Philosophy of Biological and Biomedi-*

cal Sciences 30 (1999): 273–296; Anne Marie Moulin, *Le dernier langage de la médicine: Histoire de l'immunologie de Pasteur au Sida* (Paris: Presses Universitaires de France, 1991).

3. C. von Pirquet, "Allergie," *Münchener Medizinishce Wochenschrift* 30 (1906): 1457.

4. A. Wolff-Eisner, *Das Heufiber, sein Wesen und seine Behandlung* (Munich: Lehmano, 1906).

5. S. J. Meltzer, "Bronchial Asthma as a Phenomenon of Anaphylaxis," *Journal of the American Medical Association* 55 (1910): 1021–1024. See also M. B. Emanuel, "Asthma and Anaphylaxis: A Relevant Model for Chronic Disease? An Historical Analysis of Directions in Asthma Research," *Clinical and Experimental Allergy* 25 (1995): 15–26; Mark Jackson, "John Freeman, Hay Fever, and the Origins of Clinical Allergy in Britain, 1900–1950," *Studies in History and Philosophy of Biological and Biomedical Sciences* 34 (2003): 473–490.

6. Lederle Antitoxin Laboratories, *Prophylaxis and Treatment of Hay-Fever with Pollen Vaccine* (New York: Lederle, 1916).

7. Oren C. Durham, *Your Hay Fever* (New York: Bobbs-Merrill, 1936), 145. See also Gregg Mitman, "When Pollen Became Poison: A Cultural Geography of Ragweed in America," in *The Moral Authority of Nature,* ed. Lorraine Daston and Fernando Vidal (Chicago: University of Chicago Press, 2004), 438–465; Zachary James Sopher Falck, "Controlling Urban Weeds: People, Plants, and the Ecology of American Cities, 1888–2003," (Ph.D. diss., Carnegie Mellon University, 2004).

8. W. P. Dunbar, "Observations on the Cause and Treatment of Hay Fever," *St. Louis Medical Review* 1 (1904): 177–181; C. Joachim, "The Antitoxin Treatment of Hay-Fever—Personal Observations in Prof. Dunbar's Laboratory," *New Orleans Medical and Surgical Journal* 57 (1904): 480–489; Fritzche Bros., *Etiology of Hay Fever in Picture and Word* (New York: Fritzche Bros., 1919). See also Carla C. Keirns, "Short of Breath: A Social and Intellectual History of Asthma in the United States" (Ph.D. diss., University of Pennsylvania, 2004).

9. John Freeman, "Further Observations on the Treatment of Hay Fever by Hypdermic Inoculations of Pollen Vaccine," *Lancet* 2 (1911): 814–817. See also Jackson, "John Freeman."

10. Karl K. Koessler: "The Specific Treatment of Hay Fever (Pollen Disease)," in *Forchheimer's Therapeusis of Internal Diseases,* ed. Frank Billings and Ernest E. Irons, vol. 5 (New York: D. Appleton, 1914), 703, and "The Specific Treatment of Hay Fever by Active Immunization," *Illinois Medical Journal,* August 1914: 120–127.

11. Koessler, "The Specific Treatment of Hay Fever (Pollen Disease)," 703; Ledelere Antitoxin Laboratories, *Prophylaxis and Treatment of Hay-Fever with Pollen Vaccine.*

12. Grant Selfridge, "Spasmodic Vaso-motor Disturbances of the Respiratory Tract with Special Reference to Hay Fever," *California State Journal of Medicine* 16 (1918): 170. Selfridge quotes the Lederle letter in this article.

13. W. V. Mullin, "Pollen and Hay Fever—a Regional Problem," *Colorado Medicine* 20 (1923): 99.

14. G. Piness, H. Miller, and H. E. McMinn, "Botanical Survey of Southern California in Relation to the Study of Allergic Diseases," *Bulletin of the Southern California Academy of Sciences* 25 (1926): 37.

15. Selfridge, "Spasmodic Vaso-motor Disturbances of the Respiratory Tract," 170.

16. Joseph Goodale, "Studies Regarding Anaphylactic Reactions Occurring in Horse Asthma and Allied Conditions," *Transactions of the 36th Annual Meeting of the American Laryngological Association* 171 (1914): 95–110. On Goodale, see Sheldon G. Cohen, "Firsts in Allergy, Boston Remembered," *New England and Regional Allergy Proceedings* 4 (1983): 309–334; 5 (1984): 48–64.

17. Joseph Goodale, "Pollen Therapy in Hay Fever," *Boston Medical and Surgical Journal* 173 (1915): 42–48.

18. Roger P. Wodehouse: "A Simple Method of Obtaining Ragweed Pollen in Large Quantities," *Boston Medical and Surgical Journal* 175 (1916): 430; "Preparation of Vegetable Food Proteins for Anayphylactic Tests," *Boston Medical and Surgical Journal* 175 (1916): 195–196; and "Immunochemical Studies of the Plant Proteins: Proteins of the Wheat Seed and Other Cereals: Study IX," *American Journal of Botany* 4 (1917): 417–429. On Chandler, see Cohen, "Firsts in Allergy."

19. Roger P. Wodehouse, "The Phylogenetic Value of Pollen-Grain Characters," *Annals of Botany* 42 (1928): 891–934.

20. R. Cooke and A. Vander Veer, Jr., "Human Sensitization," *Journal of Immunology* 1 (1916): 201.

21. G. Piness and H. Miller, "Specificity of Pollen Allergens," *Journal of Allergy* 1 (1930): 483–488.

22. Roger P. Wodehouse and Arthur F. Coca, "Pollen Antigens: Critical Review of Literature," *Annals of Allergy* 4 (January–February 1946): 66.

23. Ibid., 64.

24. Roger P. Wodehouse, "Pollen in Hayfever," *Torreya* 36 (1936): 77–87.

25. Wodehouse and Coca, "Pollen Antigens," 69.

26. H. M. Hall to Carlotta Hall, 10 June 1917, Box 1, HMHP.

27. H. M. Hall to Family, 11 August 1916, Box 1, HMHP.

28. H. M. Hall and F. E. Clements, *The Phylogenetic Method in Taxonomy* (Washington, D.C.: Carnegie Institute of Washington, 1923).

29. U.S. Hay Fever Association, *Forty-Third Anniversary Report* (1916).

30. Asa Gray, *Manual of the Botany of the Northern United States,* 3rd ed. (New York: Ivison, Phinney, 1862), 212.

31. Upton Sinclair, *The Jungle* (1905; New York: Signet Classic, 1980), 30.

32. Jane Addams, *Twenty Years at Hull-House* (New York: MacMillan, 1912), 99.

33. Suzanne M. Spencer-Wood, "Turn of the Century Women's Organizations, Urban Design, and the Origin of the American Playground Movement," *Landscape Journal* 13 (1994): 125–137.

34. B. Rosing, "Chicago's Unemployed Help Clean the City," *Charities and the Commons* 21 (1908): 51.

35. Robert Hessler, "Weeds and Diseases," *Survey* 26 (1911): 54, 60.

36. Durham, *Your Hay Fever,* 145–146.

37. William Scheppegrell, "Hay Fever and the National Flower," *Science* 49 (1919): 284–285; Horace Gunthorp, "Hay-Fever and a National Flower," *Science* 49 (1919): 147–148.

38. "The Rogue's Gallery of the Plant World," *Better Homes and Gardens* 12 (August 1934): 32, 51.

39. Scheppegrell, *Hayfever and Asthma: Care, Prevention, and Treatment* (Philadelphia: Lea and Febiger, 1922), 38.

40. U.S. Hay Fever Association: *Forty-second Anniversary Report* (1915), 7–8, 26, and *Forty-third Anniversary Report* (1916), 14.

41. See W. Scheppegrell to Miss L. B. Gachus, 7 August 1916, in U.S. Hay Fever Association, *Forty-fourth Anniversary Report* (1917), 24. See also William Scheppegrell: "The Seasons, Causes, and Geographical Distribution of Hay Fever and Hay Fever Resorts in the United States," *Public Health Reports* 35 (24 September 1920): 2241–2264, and *Hayfever and Asthma.*

42. William Scheppegrell: "Anaphylaxis Due to Pollen Protein, with a Report of the Results of Treatment in the Hay-Fever Clinic of the New Orleans Charity Hospital," *The Laryngoscope* 28 (1918): 859, and "Hay-Fever: Its Cause and Prevention," *Journal of the American Medical Association* 66 (4 March 1916): 707–712. In El Paso in 1927, an extensive hay fever weed eradication program was mobilized by the Women's Auxiliary of the El Paso County Medical Association. See N. S. Ives, "Weed Eradication by Community Effort Reduces Hay Fever," *American City* 37 (1927): 214. In 1938, a massive statewide "war on ragweed" in Michigan was assisted by the Boy Scouts, Camp Fire Girls, women's garden clubs, and Daughters of the American Revolution. See, e.g., R. Ray Baker: "Boy Scouts Launch County Ragweed War," *Daily News,* 21 May 1936, and "Butler Named General of Washtenaw Ragweed Army," *Daily News,* 12 June 1936.

43. Thomas J. Bassett, "Reaping on the Margins: A Century of Community Gardening in America," *Landscape* 25 (1981): 2; Allan Sutherland, "Farming Vacant City Lots," *American Monthly Review of Reviews* 31 (1905): 569. See also R. F. Powell, "Vacant Lot Gardens vs. Vagrancy," *Charities Review* 13 (1904): 25–28; Freder W. Speirs, Samuel McCune Lindsay, and Franklin B. Kirkbride, "Vacant-Lot Cultivation," *Charities Review* 8 (1898): 74–107.

44. Powell, "Vacant Lot Gardens vs. Vagrancy," 27–28.

45. "Chicago Employs 1,350 in Hay Fever Fight," *New York Times,* 12 August 1932, 17. On New York City efforts, see "Allergy, Pollen, and Ah-choo Time," *Literary Digest* 123 (12 June 1937): 17; Falck, "Controlling Urban Weeds," 166–223.

46. "Chasing Hay-Fever Pollens by Airplane," *Literary Digest* 81 (26 April 1924): 24–25.

47. R. Claude Lowdermilk, "Hay Fever," *Journal of the American Medical Association* 63 (1914): 141–142.

48. W. W. Duke and O. C. Durham: "A Botanic Survey of Kansas City, Mo., and Neighboring Rural Districts," *Journal of the American Medical Association* 82 (1924): 939–944, and "Pollen Content of the Air: Relationship to the Symptoms and Treatment of Hay-Fever, Asthma, and Eczema," *Journal of the American Medical Association* 90 (1928): 1529–1532.

49. Karl K. Koessler and O. C. Durham, "A System for an Intensive Pollen Survey, with a Report of Results in Chicago," *Journal of the American Medical Association* 86 (17 April 1926): 1204–1209.

50. O. C. Durham: "Incidence of Ragweed Pollen in United States during 1929," *Journal of the American Medical Association* 94 (14 June 1930): 1907–1911, and *Your Hay Fever,* 110; "Allergy, Pollen, and Ah-Choo Time," 17.

51. Rackemann to Spain, 9 June 1944, "Report of the Joint Committee on Standards," Folder 10, Box 434, AAAAI.

52. Sheldon Cohen, "The American Academy of Allergy: An Historical Review," *Journal of Allergy and Clinical Immunology* 64 (1979): 332–466.

53. Abbott Laboratories, *Basic Data on Pollens and Spores Designed to Assist in the Diagnosis and Treatment of Hay Fever* (Chicago: Botanical Research Department, 1946).

54. Lederle Antitoxin Laboratories, *Pollen Antigens for Prophylaxis of Hay Fever* (New York, 1925).

55. George Kent, "Sneezes on the Breezes," *American Magazine* 122 (September 1936): 74. For a history of the Dust Bowl, see Donald Worster, *Dust Bowl: The Southern Plains in the 1930s* (New York: Oxford University Press, 1979).

56. Paul Horgan, *The Return of the Weed* (New York: Harper, 1936), 3.

57. Roger P. Wodehouse, "Weeds, Waste, and Hayfever," *Natural History* 43 (1939): 162.

58. W. V. Mullin, "Pollen and Hay Fever—A Regional Problem," *Transactions of the American Academy of Ophthalmology and Oto-Laryngology* 27 (1922): 476–477.

59. Wodehouse, "Weeds, Waste, and Hayfever," 162.

60. Wodehouse, "Weeds, Waste, and Hayfever," 162; Paul B. Sears, *Deserts on the March,* 4th ed. (Norman: University of Oklahoma Press, 1980), 113.

61. Wodehouse, "Weeds, Waste, and Hayfever," 162. On the ecology of Artemisia in range management, see Hall and Clements, *The Phylogenetic*

Method in Taxonomy, 110. The notion of ecologists as land doctors is from Aldo Leopold, *A Sand County Almanac with Essays on Conservation from Round River* (New York: Ballantine Books, 1970).

62. Durham, *Your Hay Fever,* 160–161; Roger P. Wodehouse, "Hold That Sneeze!" *Rotarian* 57 (July 1940): 43–44.

63. H. A. Whittaker et al., "Report of the Committee on Air Pollution," *American Journal of Public Health* 38 (May 1948): 761–769.

64. Durham, *Your Hay Fever,* 154; Wodehouse, "Weeds, Waste, and Hayfever," 150; "Hay Fever and How to Escape It," *Popular Mechanics* 70 (July 1938): 56.

65. Whittaker et al., "Report of the Committee on Air Pollution," 768.

66. Nicolas Rasmussen, "Plant Hormones in War and Peace: Science, Industry, and Government in the Development of Herbicides in 1940s America," *Isis* 2 (2001): 291–316.

67. "Battle on Ragweed Is Seen Easing Pest," *New York Times,* 20 September 1950, 33.

68. "H-F Day," *New Yorker* 22 (24 August 1946): 17–18; Philip Gorlin, "Planning and Organizing a Ragweed Control Program," *American City* 62 (June 1947): 88–90; Israel Weinstein and Alfred Fletcher, "Essentials for the Control of Ragweed," *American Journal of Public Health* 38 (May 1948): 664–669; "Noncompetitive," *New Yorker* 26 (August 1950): 17–18.

69. "Weinstein Defends Drive on Ragweed, Disputes Criticism by Medical Journal," *New York Times,* 17 September 1946, 17.

70. Philip Gorlin, "Ragweed Control in New York City," *Hay Fever Bulletin* 6, no. 1 (1955): 4–6; Alfred H. Fletcher, "Ragweed Control in New Jersey," *Hay Fever Bulletin* 6, no. 2 (Summer 1955): 7–8, 12.

71. HB 283, 1957, pp. 1–2, College of Agricultural Sciences Records (RG 158), II. Administration: Experiment Stations: Request for Research: Ragweed (1955–1958), OSUA; Harold M. Erickson, "Ragweed in Oregon," n.d., College of Agricultural Sciences Records (RG 158), II. Administration: Experiment Stations: Request for Research: Ragweed (1955–1958), OSUA. On landowner concerns, see Oregon State Department of Agriculture, *Ragweed Control: Report for 1958* (Salem: State Department of Agriculture, 1958), 4, and William G. Robbins, *Landscapes of Conflict: The Oregon Story, 1940–2000* (Seattle: University of Washington Press, 2004).

72. Matthew Walzer and Bernard B. Siegel, "The Effectiveness of the Ragweed Eradication Campaigns in New York City: A 9-Year Study (1946–1954)," *Journal of Allergy* 27 (1956): 113–126.

73. Rachel Carson, *Silent Spring* (1962; New York: Fawcett Crest, 1964), 78, 169, 23, 261; Edmund Russell, *War and Nature: Fighting Humans and Insects with Chemical from World War I to Silent Spring* (Cambridge: Cambridge University Press, 2001).

74. http://www.wellmark.com/health_improvement/reports/antihis tamines/about_antihistamines.htm.

75. S. Heather Duncan, "Studies: Warming Trend Could Hurt Asthma, Allergy Sufferers," *Macon Telegraph,* 13 June 2005; Lewis Ziska et al., "Cities as Harbingers of Climate Change: Common Ragweed, Urbanization, and Public Health," *Journal of Allergy and Clinical Immunology* 111 (February 2003): 290–295; Paul R. Epstein and Christine Rogers, *Inside the Greenhouse: The Impacts of* CO_2 *and Climate Change on Public Health in the Inner City* (Cambridge: Harvard Medical School, Center for Health and the Global Environment, 2004).

CHAPTER 3: THE LAST RESORTS

1. "EDF/ADHS/Asthmatics, SWES, Tucson, 12-05-85, Folder: Environmental Defense Fund_Asthmatics, Box 6, GASP.

2. EDF/ADHS/Asthmatics, p. 4, GASP.

3. "Suit Says EPA Allows Acid Rain," *Tucson Citizen,* 5 December 1985. For figures on the number of asthmatics, see "Petition to Establish Operating Conditions on Smelters in Arizona to Limit SO_2 Concentrations to Prevent Endangerment of Health of Asthmatics," 26 August 1985, Folder: Environmental Defense Fund—Asthmatics, Box 6, GASP. On smelter emission figures, see "The Smelter Triangle: A Policy Overview," 28 June 1985, Box 3, GASP.

4. Statement of Governor Bruce Babbitt of Arizona before the Health and Environment Subcommittee of the Committee on Energy and Commerce, 28 June 1985, Folder: Governor Babbitt, Box 1, GASP.

5. "Petition to Establish Operating Conditions," 1.

6. On the history of Phelps Dodge, see Carlos A. Schwantes, *Vision and Enterprise: Exploring the History of Phelps Dodge Corporation* (Tucson: University of Arizona Press, 2000).

7. "Press Statement, GASP and Earth First!" 12 August 1986, Box 3, GASP. For other newspaper coverage, see Dick Kamp, "Consent Decree a Major Step, but Bears Watching," *Bisbee Daily Review,* 24 August 1986; "The Largest SO_2 Emitter in the U.S. Is Back in Business," *High Country News,* 18 August 1986.

8. "Sulfur in the Skies," *Denver Post,* 8 March 1985.

9. Economic figures regarding Denver's early development are from Denver and Rio Grande Railway, *Health, Wealth and Pleasure, Colorado and New Mexico* (Chicago: Belford, Clarke, 1881); Lyle W. Dorsett, *The Queen City: A History of Denver* (Boulder, Colo.: Pruett Publishing, 1977); Stephen J. Leonard and Thomas J. Noel, *Denver: Mining Camp to Metropolis* (Niwot, Colo.: University Press of Colorado, 1990).

10. Estimates are from Susan Jane Edwards, "Nature as Healer: Denver, Colorado's Social and Built Landscapes of Health, 1880–1930" (Ph.D. diss., University of Colorado, 1994); Billy M. Jones, *Health-Seekers in the Southwest, 1817–1900* (Norman: University of Oklahoma Press, 1967); Sheila M. Roth-

man, *Living in the Shadow of Death: Tuberculosis and the Social Experience of Illness in American History* (New York: Basic Books, 1994).

11. Denver and Rio Grande Railway, *Health, Wealth, and Pleasure in Colorado and New Mexico* (Chicago: Belford, Clarke, 1881), 124.

12. Editorials in *Rocky Mountain News,* 2 March 1880, and 11 August 1880.

13. *To The Rockies and Beyond, or a Summer on the Union Pacific Railroad and Branches* (Chicago: Belford, Clarke, 1881), 67.

14. Grace Greenwood, *New Life in New Lands: Notes of Travel* (New York: J. B. Ford, 1873), 95.

15. H. H., "Mountain Life: The New Hampshire Town of Bethlehem," *New York Evening Post,* 18 October 1865. Although Sheila Rothman in *Living in the Shadow of Death* links Jackson's travels in search of health to the specter of consumption, Jackson's nineteenth-century biographer identified her as a hay fever sufferer, a diagnosis that fits with the places to which she traveled and their seasonality. See Thomas Wentworth Higginson, "Mrs. Helen Hunt Jackson," *Century Magazine* 31 (December 1885): 251–259. On Jackson's literary life, see Ruth Odell, *Helen Hunt Jackson* (New York: D. Appleton-Century, 1939); Kate Phillips, *Helen Hunt Jackson: A Literary Life* (Berkeley: University of California Press, 2003).

16. Helen Hunt Jackson, *Bits of Travel at Home* (Boston: Little, Brown, 1898), 191–195; H. H., "In the White Mountains," *New York Independent,* 13 September 1866, 2.

17. [Rip Van Winkel], "A Protest against the Spread of Civilization," *New York Evening Post,* 29 August 1867.

18. Jackson, *Bits of Travel at Home,* 40.

19. Ibid., 163.

20. Quoted in Phillips, *Helen Hunt Jackson,* 168.

21. Jackson, *Bits of Travel at Home,* 224.

22. S. Edwin Solly, *The Health Resorts of Colorado Springs and Manitou* (Colorado Springs: Gazette Publishing, 1883), 29.

23. Odell, *Helen Hunt Jackson,* 130.

24. Quoted in Phillips, *Helen Hunt Jackson,* 169.

25. Odell, *Helen Hunt Jackson,* 131.

26. Jackson, *Bits of Travel at Home,* 231.

27. F. J. B. Crane et al., *Colorado and Asthma* (Denver: Rocky Mountain News Steam, 1874), 15.

28. Greenwood, *New Life in New Lands,* 75, 96.

29. Kathleen A. Brosnan, *Uniting Mountain and Plain: Cities, Law, and Environmental Change along the Front Range* (Albuquerque: University of New Mexico Press, 2002).

30. Helen Hunt Jackson, "Alamosa," *New York Independent,* 6 and 13 June 1878.

31. Jackson, *Bits of Travel at Home,* 286.

32. Ibid., 384.

33. Ibid., 226.

34. *Health, Wealth, and Pleasure, Colorado and New Mexico,* 19.

35. *Denver Times,* 2 July 1889.

36. Dorsett, *The Queen City,* 126–127.

37. David McCullough, *Mornings on Horseback* (New York: Simon and Schuster, 1981), 340.

38. "Barfield Sanatorium" (Tucson, 1931), UASC.

39. C. L. Sonnichsen, *Tucson: The Life and Times of an American City* (Norman: University of Oklahoma Press, 1982); Michael F. Logan, *The Lessening Stream: An Environmental History of the Santa Cruz River* (Tucson: University of Arizona Press, 2002).

40. Michael F. Logan, *Fighting Sprawl and City Hall: Resistance to Urban Growth in the Southwest* (Tucson: University of Arizona Press, 1995); Margaret T. Parker, "Tucson: City of Sunshine," *Economic Geography* 24 (1948): 79–113.

41. Alex Jay Kimmelman, "Luring the Tourist to Tucson: Civic Promotion during the 1920s," *Journal of Arizona History* 28 (1987): 135–154; Roy P. Drachman, *From Cowtown to Desert Metropolis: Ninety Years of Arizona Memories* (San Francisco: Whitewing Press, 1999).

42. *Treasures of Health: The Special Climatic Advantages of Tucson, Arizona* (Tucson Board of Trade, 1898), 6, 10.

43. John A. Black, *Arizona: The Land of Sunshine and Silver, Health and Prosperity, The Place for Ideal Homes* (Phoenix: Republican Book and Job Print, 1890), 21.

44. Kimmelman, "Luring the Tourist to Tucson," 137; Tucson Sunshine Climate Club, *Man-Building in the Sunshine Climate* (Tucson, 1923).

45. Parker, "Tucson," 103–106.

46. Lawrence V. Tagg, *Harold Bell Wright: Storyteller to America* (Tucson: Westernlore Press, 1986), 79.

47. Harold Bell Wright, "Why I Did Not Die," *American Magazine* 97 (June 1924): 15.

48. Harold Bell Wright, *The Mine with the Iron Door* (New York: D. Appleton, 1923), 4. For more on Wright's influence in Tucson, see James L. Sell, "The Novelist Who Shaped the City," *Tucson Weekly,* 9 November 2000.

49. Wright, *The Mine with the Iron Door,* 28–29.

50. Wright, "Places of Healing," *Outdoor America* 6 (1928): 4; Tagg, *Harold Bell Wright,* 152.

51. Wright, *The Mine with the Iron Door,* 167.

52. Sell, "The Novelist Who Shaped the City."

53. Quoted in *Treasures of Health,* 7.

54. Dick Hall, "Ointment of Love: Oliver E. Comstock and Tucson's Tent City," *Journal of Arizona History* 19 (1978): 111–130.

55. Parker, "Tucson," 103.

56. "Sanatoria and Rest Homes in Tucson," August 1936; Vertical File: Sanitariums Printed Material, Folder 1, UASC.

57. Reba Douglass Grub, *Portrait of Progress: A Story of Tucson Medical Center* (Tucson: Raim, 1984). On the Desert Botanical Laboratory, see Sharon E. Kingsland, *The Evolution of American Biology, 1890–2000* (Baltimore: Johns Hopkins University Press, 2005), 96–128.

58. Desert Sanatorium, *The Desert Sanatorium and Institute of Research* (Tucson: Desert Sanatorium, 1932), 10 13.

59. Grub, *Portrait of Progress,* 22–29.

60. Untitled proposal, Box 3, "Box 7," Miscellany, JMP. See also additional correspondence between MacDougal and Metzger in this file.

61. "Pupin on Staff of Sanatorium," *Arizona Daily Star,* 14 April 1929; *Treasures of Health,* 7.

62. Ibid.

63. Grub, *Portrait of Progress,* 36–50; Sonnichsen, *Tucson,* 272.

64. See, e.g., Barbara Bates, *Bargaining for Life: A Social History of Tuberculosis, 1876–1938* (Philadelphia: University of Pennsylvania Press, 1992); Mark Caldwell, *The Last Crusade: The War on Consumption, 1862–1954* (New York: Atheneum, 1988); Rothman, *Living in the Shadow of Death.*

65. The 2.5 million is a rough estimate taken from M. Murray Peshkin, "From a Home to a Hospital," 1 August 1955, Folder 7, Box 10, NAC.

66. "Tucson to Be Child Asthma Treatment Center," *Arizona Daily Star,* 28 October 1951; W. B. Steen, "Rehabilitation of Children with Intractable Asthma," *Annals of Allergy* 17 (1959): 871.

67. "Agenda: Special Board Meeting," 19 August 1956, Box 2, Minutes, 1955 and 1956 Folder, NAC.

68. See Jane S. Smith, *Patenting the Sun: Polio and the Salk Vaccine* (New York: William Morrow, 1990). On the history of poliomyletis in the United States, see Naomi Rogers, *Dirt and Disease: Polio before FDR* (New Brunswick, N.J.: Rutgers University Press, 1992).

69. "National Fund Drive Opens for Asthmatic Youths," *Arizona Daily Star,* 13 January 1953; "Asthmatic Foundation Launches Fund Drive," *Arizona Daily Star,* 30 October 1958; "Asthma Facility Ground Broken," *Arizona Daily Star,* 30 November 1960.

70. See Gregg Mitman, "Geographies of Hope: Mining the Frontiers of Health in Denver and Beyond, 1870–1965," *Osiris* 19 (2004): 93–111.

71. National Foundation for Asthmatic Children, *That They May Breathe* (Tucson: NFAC, n.d.).

72. Peshkin, "From a Home to a Hospital," 2.

73. M. Murray Peshkin, "Survey of Convalescent Institutions for Asthmatic Children in United States and Canada," *Journal of Asthmatic Research* 2 (1965): 181–194.

74. M. Murray Peshkin, "Asthma in Children: IX. Role of Environment

in Treatment of a Selected Group of Cases: A Plea for a 'Home' as a Restorative Measure," *American Journal of Diseases of Children* 39 (1930): 776.

75. M. Murray Peshkin, "The Child Rescue Work of the National Home for Jewish Children," Box 11, Histories, 1929–83 Folder, NAC.

76. John E. Allen, "How We Help Asthmatic Children," *Saturday Evening Post,* 3 September 1960, 54; "National Fund Drive Opens for Asthmatic Youths."

77. Presidential Proclamation 3777, "National CARIH Asthma Week," F.R. Doc. 67-1421. NIH budget statistics are available in the *NIH Almanac* at http://www.nih.gov/about/almanac/index.html.

78. "Jewish National Home for Asthmatic Children at Denver," Box 11, Description of Programs, 1937 Folder, NAC.

79. Evelyn Walsh, "The Aims of a Medical Laboratory," 9 March 1949, Box 1, Minutes of Board, 1948 and 1949 Folder, NAC.

80. See, e.g., "Annual Report of the Medical Director," 1954–1955, Box 2, Minutes, 1955 and 1956 Folder, NAC; "News from the Home Front," September 1952, Box 9, Folder 3, NAC.

81. On the U.S. Public Health Service grant, see "President's Annual Report," 21 January 1959, Box 2, Minutes 1957–58–59 Folder, NAC. For income statistics, see Executive Director's Report, 1972, Box 11, Jack Gershterson Report, 1972 Folder, NAC.

82. "Climate Is Key to Asthma Relief," *Arizona Daily Star,* 27 March 1953.

83. "Sahuaro School Provides New Hope for Asthmatics," *Arizona Daily Star,* 1 December 1963; "Asthmatic Project Aided," *Arizona Daily Star,* 10 January 1962.

84. "Fund Earmarked for UA Research," *Arizona Daily Star,* 9 May 1965.

85. "Increasing Costs Doom Asthmatics' School, *Arizona Daily Star,* 23 March 1969.

86. Scott Hamilton Dewey, *Don't Breathe the Air: Air Pollution and U.S. Environmental Politics,* 1945–1970 (College Station: Texas A and M University Press, 2000).

87. J. J. Schueneman, *The Denver Air Pollution Problem* (Cincinnati: U.S. Public Health Service, 1957). See also Leonard A. Dobler and Joseph Palomba, *Denver Metropolitan Area Air Pollution Emission Inventory* (Denver, 1963); Air Pollution Control Division, *Assessment of Air Quality: Metro Denver Region* (Denver: Air Pollution Control Division, 1975); John G. Watson et al., *The 1987–88 Metro Denver Brown Cloud Study,* DRI Document No. 8810 1F1 (Denver, 1988).

88. Logan, *Fighting Sprawl and City Hall.*

89. Sonnichsen, *Tucson,* 285.

90. Michael D. Lebowitz and Benjamin Burrows, "Tucson Epidemiologic Study of Obstructive Lung Diseases, II: Effects of In-Migration Factors on the Prevalence of Obstructive Lung Diseases," *American Journal of Epidemiology* 102 (1975): 153–163.

91. "Insinuation That City Has Smog Easily Refuted," *Arizona Daily Star,* 24 January 1957.

92. John D. Margolis, *Joseph Wood Krutch: A Writer's Life* (Knoxville: University of Tennessee Press, 1980), xiv.

93. Joseph Wood Krutch, *The Desert Year* (New York: William Sloane Associates, 1951), 184.

94. Joseph Wood Krutch, *If You Don't Mind My Saying So . . .* (New York: William Sloane Associates, 1964), 361–362.

95. *Tucson's Healthful Climate Will Relieve Pain and Suffering for Most People with Arthritis, Sinusitis, Rheumatism, Asthma, Bronchitis, Emphysema, and Cardio-Vascular Disease* (Tucson's Healthful Climate Association, n.d.).

96. "Smog Layer Gives City Its Worst Air Pollution in Two-Year Period," *Arizona Daily Star,* 19 January 1961; "Smog Problem under Survey," *Arizona Daily Star,* 7 February 1961.

97. Quentin M. Mees and Robert L. Worman, "Preliminary Report, Air Pollution Surveillance Study, Tucson, Arizona"; Arizona State Department of Health, Engineering Experiment Station, Bull. No. 13, Civil Engineering Series No. 6, Project No. 59-CE-2, 23.

98. Quentin M. Mees, *Tucson Air Pollution, 1959–1964,* EES Series Report No. 7 (Tucson: University of Arizona, 1964), 36.

99. M. D. Lebowitz et al., "The Effect of Air Pollution and Weather on Lung Function in Exercising Children and Adolescents," *American Review of Respiratory Diseases* 109 (1974): 262.

100. Samuel H. Watson and Charles S. Kibler, "Etiology of Hay-Fever in Arizona and the Southwest," *Journal of the American Medical Association* 78 (1922): 719–722; E. W. Phillips, "The Changing Pollen Picture," *Southwestern Medicine* 16 (1932): 70–74.

101. Melvin E. Hecht, "The Decline of the Grass Lawn Tradition in Tucson," *Landscape* 19 (1975): 3–10; E. Gregory McPherson and Renee A. Haip, "Emerging Desert Landscape in Tucson," *Geographical Review* 79 (1989): 435–449.

102. Mark R. Sneller, Harry D. Hayes, and Jacob L. Pinnas, "Pollen Changes during Five Decades of Urbanization in Tucson, Arizona," *Annals of Allergy* 71 (1993): 519–524.

103. Martin Robert Yoklic, "Airborne Pollen and Tucson's Urban Landscape: A Model for Policy Determination" (master's thesis in architecture, University of Arizona, 1983); Mark R. Sneller, Harry D. Hayes, and Jacob L. Pinnas, "Frequency of Airborne Alternaria Spores in Tucson, Arizona over a 20-Year Period," *Annals of Allergy* 46 (1981): 30–33; Marilyn Holonen et al., "*Alternaria* as a Major Allergen for Asthma in Children Raised in a Desert Environment," *American Journal of Respiratory and Critical Care Medicine* 155 (1997): 1356–1361.

104. Allen M. Solomon and Harry D. Hayes, "Impacts of Urban Development upon Allergenic Pollen in a Desert City," *Journal of Arid Environments*

3 (1980): 169–178; Michael D. Lebowitz, Ronald J. Knudson, and Benjamin Burrows, "Tucson Epidemiologic Study of Obstructive Lung Diseases, I: Methodology and Prevalence of Disease," *American Journal of Epidemiology* 102 (1975): 137–151.

105. "City Ponders Pollen Ban: Mulberry Not the Only Villain, *Arizona Daily Star,* 3 October 1975.

106. Hecht, "The Decline of the Grass Lawn Tradition," 7.

107. Patient to Mr. G, Box 14, Alumni Correspondence, 1980, NAC.

108. http://www.ci.tucson.az.us/about.html.

CHAPTER 4: CHOKING CITIES

1. Gordon Parks, "Freedom's Fearful Foe: Poverty," *Life,* 16 June 1961, 86–98.

2. "Flavio's Rescue: Americans Bring Him from Rio Slum to Be Cured," *Life,* 21 July 1961, 24.

3. Gordon Parks, *Flavio* (New York: W. W. Norton, 1978), 91–92.

4. Piri Thomas, *Down These Mean Streets* (New York: Vintage, 1974), xi.

5. W. Carr, L. Zeitel, and K. Weiss, "Variations in Asthma Hospitalizations and Deaths in New York City," *American Journal of Public Health* 82 (1992): 59–65; V. A. De Palo, P. H. Mayo, P. Friedman et al., "Demographic Influences on Asthma Hospitalization Rates in New York City," *Chest* 106 (1994): 447–451; L. Stevenson and M. Kaminsky, "Asthma Hospitalization and Mortality in New York City, 1987–1996," in *Conference Proceedings: Working Together to Combat Urban Asthma* (New York: New York Academy of Medicine, 1998).

6. Kenneth B. Clark, *Dark Ghetto: Dilemmas of Social Power* (New York: Harper and Row, 1965), 11.

7. Raymond Arsenault, *Freedom Riders: 1961 and the Struggle for Racial Justice* (Oxford: Oxford University Press, 2006).

8. "Orleans Asthma Outbreak Noted," *Times-Picayune,* 19 October 1961, 1.21.

9. On the racial geography of New Orleans, see Pierce F. Lewis, *New Orleans: The Making of an Urban Landscape,* 2nd ed. (Sante Fe, N.M.: Center for American Places, 2003). On the economic conditions of blacks in New Orleans during this period, see Forrest E. Violate, with the assistance of Joseph T. Taylor and Giles Hubert, "The Negro in New Orleans," in *Studies in Housing and Minority Groups,* ed. Nathan Glazer and Davis McEntire (Berkeley: University of California Press, 1960), 135–143. On Charity Hospital, see John Salvaggio, *New Orleans' Charity Hospital: A Story of Physicians, Politics and Poverty* (Baton Rouge: Louisiana State University Press, 1992).

10. John D. Walsh and Vincent J. Derbes, "The Fifth Season: New Orleans Asthma," *NTA Bulletin,* December 1964, 8–9.

11. Robert F. Lewis and Edward A. Cleave, *A Study of Relationships between*

Meteorologic Variables and Asthma Admissions to Charity Emergency Center, New Orleans, Louisiana (New Orleans: Tulane University, 1960?); Robert F. Lewis, Murray M. Gilkeson, Jr., and Ray W. Robison, *Air Pollution and New Orleans Asthma,* 2 vols. (New Orleans: Tulane University, 1962); "A City Chokes Up," *Newsweek,* 14 November 1960, 104.

12. Quoted in Devra Davis, *When Smoke Ran like Water: Tales of Environmental Deception and the Battle against Pollution* (New York: Basic Books, 2002), 16; Lynn Snyder, " 'The Death-Dealing Smog over Donora, Pennsylvania': Industrial Air Pollution, Public Health, and Federal Policy, 1915–1963" (Ph.D., diss., University of Pennsylvania, 1994).

13. Robert B. W. Smith et al., "Tokyo-Yokohama Asthma: An Area Specific Air Pollution Disease," *Archives of Environmental Health* 8 (1964): 805–817.

14. Robert F. Lewis, Murray M. Gilkeson, and Roy O. McCaldin, "Air Pollution and New Orleans Asthma," *Public Health Reports* 77 (1962): 947–954.

15. Quoted in Amanda Furness, "Officials and Activists Join Forces to Clean Up Dump Sites," *Louisiana Weekly,* 8 October 2001.

16. For a history of the Agriculture Street landfill, see Louisiana Office of Public Health, *Public Health Assessment. Agriculture Street Landfill. New Orleans, Orleans Parish, Louisiana. EPA Facility ID: LAD981056997,* 2 June 1999.

17. Robert F. Lewis, "Epidemic Asthma in New Orleans: A Summary of Knowledge to Date," *Journal of the Louisiana State Medical Society* 115 (1963): 300–303; Hans Weill et al.: "Further Observations on New Orleans Asthma," *Archives of Environmental Health* 8 (1964): 192–195, and "Epidemic Asthma in New Orleans," *Journal of the American Medical Association* 190 (1964): 311–314.

18. Hans Weill, Morton M. Ziskind, Richard C. Dickerson, and Vincent J. Derbes, "Allergenic Air Pollutants in New Orleans," *Journal of the Air Pollution Control Association* 15 (1964): 467–471.

19. John Salvaggio, Lawrence Zaslow, John Greer, and John Seabur, "New Orleans Asthma, III: Semiquantitative Aerometric Pollen Sampling, 1967 and 1968," *Annals of Allergy* 29 (1971): 305–317.

20. Craig E. Colten, "Basin Street Blues: Drainage and Environmental Equity in New Orleans, 1890–1930," *Journal of Historical Geography* 28 (2002): 237–257.

21. John Salvaggio, John Seabur, and Enno A. Schoenhardt, "New Orleans Asthma, V: Relationships between Charity Hospital Asthma Admission Rates, Semiquantitative Pollen and Fungal Spore Counts, and Total Particulate Aerometric Sampling Data," *Journal of Allergy and Clinical Immunology* 48 (1971): 113.

22. Scott Hamilton Dewey, *Don't Breathe the Air: Air Pollution and U.S. Environmental Politics, 1945–1970* (College Station: Texas A and M University Press, 2000). "Smog Is Really Smaze: Rain May Rout It Tonight," *New York*

Times, 21 November 1953, 1; "Rain and Wind Lift City's 6-Day Siege of Smaze and Smog," *New York Times,* 23 November 1953, 1.

23. Leonard Greenburg et al.: "Air Pollution and Morbidity in New York City," *Journal of the American Medical Association* 182 (1962): 161–164, and "Report of an Air Pollution Incident in New York City, November 1953," *Public Health Reports* 77 (1962): 7–16.

24. Leonard Greenburg, "Air Pollution and Asthma," *Journal of Asthma Research* 2 (1965): 195–198.

25. Leonard Greenburg, Carl Erhardt, Franklyn Field, and Joseph I. Reed, "Air Pollution Incidents and Morbidity Studies," *Archives of Environmental Health* 10 (1965): 351–356; Michael Earnest et al., "Emergency Clinic Visits for Asthma," *Public Health Reports* 81 (1966): 911–918.

26. Clark, *Dark Ghetto,* 22; Douglas S. Massey and Nancy A. Denton, *American Apartheid: Segregation and the Making of the Underclass* (Cambridge: Harvard University Press, 1993); Steward E. Tolnay, "The Great Migration and Changes in the Northern Black Family, 1940 to 1990," *Social Forces* 75 (1997): 1213–1238.

27. These statistics are taken from HARYOU, *Youth in the Ghetto: A Study of the Consequences of Powerlessness and a Blueprint for Change* (New York: Orans Press, 1964). Quote appears on p. 316.

28. Ibid.; Richard Pomeroy, in collaboration with Robert LeJeune and Lawrence Podell, *Studies in the Use of Health Services by Families on Welfare: Utilization by Publicly-Assisted Families* (New York: Center for the Study of Urban Problems, 1969), 114.

29. Virginia E. Sánchez Korrol, *From Colonia to Community: The History of Puerto Ricans in New York City,* rev. ed. (Berkeley: University of California Press, 1994).

30. Quoted in Thomas, *Down These Mean Streets,* 10.

31. Beatrice Bishop Berle, *80 Puerto Rican Families in New York City: Health and Disease Studied in Context* (New York: Columbia University Press, 1958). See also Elena Padilla, *Up from Puerto Rico* (New York: Columbia University Press, 1958); Patricia Cayo Sexton, *Spanish Harlem: Anatomy of Poverty* (New York: Harper and Row, 1965).

32. "Race Tension Shows Contrasting Pattern in North and South," *New York Times,* 26 July 1964, E3; E. W. Kenworthy, "Johnson Orders Full F.B.I. Inquiry in Harlem Riots," *New York Times,* 22 July 1964, 1; Paul L. Montgomery and Francis X. Clines, "Thousands Riot in Harlem Area; Scores Are Hurt," *New York Times,* 19 July 1964, 1.

33. John A. Osmundsen, "Asthma Linked to Rights Drive," *New York Times,* 26 July 1965, 25.

34. John A. Osmundsen, "Asthma Linked to Emotions," *New York Times,* 1 August 1965, E7.

35. See, e.g., Michael Hirt, ed., *Psychological and Allergic Aspects of Asthma* (Springfield, Ill.: Charles C. Thomas, 1965).

36. Abram Kardiner and Lionel Ovesey, *The Mark of Oppression: A Psychosocial Study of the American Negro* (New York: W. W. Norton, 1951), 310. On the history of the damaged black personality, see Daryl Michael Scott, *Contempt and Pity: Social Policy and the Image of the Damaged Black Psyche, 1880–1996* (Chapel Hill: University of North Carolina Press, 1997).

37. Osmundsen, "Asthma Linked to Rights Drive," 25.

38. Samuel C. Bukantz: "Initial Medical Evaluation Conferences," *Journal of Children's Asthma Research Institute* 1 (March 1961): 132, and "Final and Interim Medical Evaluation Conferences," *Journal of the Children's Asthma Research Institute and Hospital* 1 (June 1961): 173–204.

39. On white privilege and imperception, see Michelle Murphy, "Uncertain Exposures and the Privilege of Imperception: Activist Scientists and Race at the U.S. Environmental Protection Agency," *Osiris* 19 (2004): 266–282.

40. Harry S. Bernton and Halla Brown, "Insect Allergy: Preliminary Studies of the Cockroach," *Journal of Allergy* 35 (1964): 506–513. See also Richard J. Brenner et al., "Modernized Society and Allergies to Arthropods," *American Entomologist* 37 (1991): 143–155.

41. Harry S. Bernton and Halla Brown, "Cockroach Allergy II: The Relation of Infestation to Sensitization," *Southern Medical Journal* 60 (1967): 852–855.

42. Jane E. Brody, "Allergy to Cockroaches Called One Possible Cause of Asthma," *New York Times,* 2 September 1967, 15.

43. Robert W. Stock, "It's Always the Year of the Roach," *New York Times Sunday Magazine,* 21 January 1968, 34–39.

44. *Cockroaches: How to Control Them* (U.S. Department of Agriculture, 1969).

45. Hugo Hartnack, *202 Common Household Pests* (Chicago: Hartnack Publishing, 1939), 86.

46. *West Side Story* (1961). For cultural histories of the cockroach, see Marion Copeland, *Cockroach* (London: Reaktion Books, 2003), and Richard Schweid, *The Cockroach Papers: A Compendium of History and Lore* (New York: Four Walls Eight Windows, 1999).

47. Michael K. Rust, John M. Owens, and Donald A. Reierson, eds., *Understanding and Controlling the German Cockroach* (New York: Oxford University Press, 1995); William J. Bell and K. G. Adiyodi, *The American Cockroach* (New York: Chapman and Hall, 1982).

48. Brenner et al., "Modernized Society and Allergies to Arthropods"; Angel M. Marchand, "Allergy to Cockroaches," *Boletín de la Asociación Médica de Puerto Rico,* February 1966, 49–53.

49. R. C. Akers and W. H. Robinson, "Spatial Pattern and Movement of German Cockroaches in Urban, Low-Income Apartments (Dictyoptera: Blattellidae)," *Proceedings of the Entomological Society of Washington* 83 (1981): 168–172; Erik S. Runstrom and Gary W. Bennett, "Movement of German Cockroaches (Orthoptera: Blattellidae) as Influenced by Structural Features

of Low-Income Apartments," *Journal of Economic Entomology* 77 (1984): 407–411.

50. Gordon Parks, "The Cycle of Despair," *Life,* 8 March 1968, 50.

51. Warren Miller, *The Cool World* (New York: Fawcett Premier, 1959), 26.

52. Michael Katz, *The Undeserving Poor: From the War on Poverty to the War on Welfare* (New York: Pantheon Books, 1989).

53. Richard A. Cloward, "The War on Poverty: Are the Poor Left Out?" *The Nation,* 2 August 1965, 55.

54. Herbert Krosney, "Mobilization for Youth: Feuding over Poverty," *The Nation,* 14 December 1964, 455–461.

55. Cloward, "The War on Poverty," 59.

56. HARYOU, *Youth in the Ghetto.* On the history of HARYOU and community empowerment in Harlem, see Gerald Markowitz and David Rosner, *Children, Race, and Power: Kenneth and Mamie Clark's Northside Center* (New York: Routledge, 2000), 180–216.

57. Ibid., 567–568.

58. "Harlem Tenants Open Rent Strike," *New York Times,* 28 September 1963, 44.

59. "Tenants in 34 Tenements Join Growing Rent Strike in Harlem," *New York Times,* 2 December 1963, 30; "Slum Rent Strike Gains Momentum," *New York Times,* 1 January 1964, 28.

60. "City Plans Drive on Rats in Slums," *New York Times,* 11 January 1964, 1.

61. "Rent-Strike Chief and 10 Arrested," *New York Times,* 8 February 1964, 1.

62. "200 Rubber Rats Sent to Governor," *New York Times,* 16 February 1964.

63. "1,000 in Harlem Cheer Malcolm X," *New York Times,* 23 March 1964, 18.

64. Jose Yglesias, "Right on with the Young Lords," *New York Times,* 7 June 1970, 86.

65. For an excellent account of the Young Lords in relation to environmental justice, see Matthew Gandy, *Concrete and Clay: Reworking Nature in New York City* (Cambridge: MIT Press, 2002), 162–186.

66. Yglesias, "Right on with the Young Lords"; Gandy, *Concrete and Clay.*

67. I. F. Goldstein and John Salvaggio, "The Decline of New Orleans Asthma Epidemics," *Reviews on Environmental Health* 4 (1984): 133–146.

68. Katz, *The Undeserving Poor,* 189–190.

69. Rodrick Wallace and Deborah Wallace, "Origins of Public Health Collapse in New York City: The Dynamics of Planned Shrinkage, Contagious Urban Decay, and Social Disintegration," *Bulletin of the New York Academy of Medicine* 66 (1990): 391–434.

70. R. Allen Hayes, *The Federal Government and Urban Housing: Ideology and Change in Public Policy* (Albany: State University of New York Press,

1995); James O. Wilson, ed., *Urban Renewal: The Record and the Controversy* (Cambridge: MIT Press, 1966).

71. Harlem Task Force, *A Profile of the Harlem Area* (December 1973), SC-NYPL.

72. Alexander von Hoffman, "High Ambitions: The Past and Future of American Low-Income Housing Policy," *Housing Policy Debate* 7 (1996): 423–446.

73. Asthma—United States, 1980–1987," *Morbidity and Mortality Weekly Report* 30 (1990): 493–497; D. Mannino et al.: "Surveillance for Asthma Prevalence—United States, 1960–1995," *Morbidity and Mortality Weekly Report* 47 (1998): 1–27, and "Surveillance for Asthma—United States, 1980–1999," *Morbidity and Mortality Weekly Report* 51 (2002): 1–13. See also Kevin B. Weiss, Peter J. Gergen, and Ellen F. Crain, "Inner-City Asthma: The Epidemiology of an Emerging U.S. Public Health Concern," *Chest* 101 (June 1992): 362S–367S.

74. W. Carr, L. Zeitel, and K. Weiss, "Asthma Hospitalization and Mortality in New York City," *American Journal of Public Health* 82 (1992): 59–65.

75. Brian P. Leaderer et al., "Dust Mite, Cockroach, Cat, and Dog Allergen Concentrations in Homes of Asthmatic Children in the Northeastern United States: Impacts of Socioeconomic Factors and Population Density," *Environmental Health Perspectives* 110 (2002): 419–425.

76. Stephen W. Nicholas, "Addressing the Childhood Asthma Crisis in Harlem: The Harlem Children's Zone Asthma Initiative," *American Journal of Public Health* 95 (2005): 245–248.

77. Lawrence W. Wissow et al., "Poverty, Race, and Hospitalization for Childhood Asthma," *American Journal of Public Health* 78 (1988): 777–782; D. M. Lang and M. Polansky, "Patterns of Asthma Mortality in Philadelphia from 1969 to 1991," *New England Journal of Medicine* 331 (1994): 1542–1546.

78. See, for example, David Rosenstreich et al., "The Role of Cockroach Allergy and Exposure to Cockroach Allergen in Causing Morbidity among Inner-City Children with Asthma," *New England Journal of Medicine* 336 (1997): 1356–1363; Jianfeng Zu et al., "Genomewide Screen and Identification of Gene-Gene Interactions for Asthma Susceptibility Loci in Three U.S. Populations: Collaborative Study on Genetics of Asthma," *American Journal of Human Genetics* 68 (2001): 1437–1446. For an overview of the National Cooperative Inner-City Asthma Study, see http://www.niaid.nih.gov/fact sheets/asthma.htm. For information on the Collaborative Study on the Genetics of Asthma, see http://www.csga.org.

79. Quoted in Phil Brown et al., "The Health Politics of Asthma: Environmental Justice and Collective Illness Experience in the United States," *Social Science and Medicine* 57 (2003): 461.

80. http://weact.org.

81. Peggy M. Shepard, "Issues of Community Empowerment," *Fordham Urban Law Journal* 21 (1994): 739. The literature on the history of environ-

mental justice in America is rapidly growing. For a sampling, see Robert Bullard, *Dumping on Dixie: Race, Class, and Environmental Quality* (Boulder, Colo.: Westview, 1990); Giovanna Di Chiro, "Nature as Community: The Convergence of Environment and Social Justice," in *Uncommon Ground: Toward Reinventing Nature,* ed. William Cronon (New York: W. W. Norton, 1995), 298–321; Michael Egan, "Subaltern Environmentalism in the United States: A Historiographic Review," *Environment and History* 8 (2002): 21–41; Gandy, *Concrete and Clay;* Robert Gottlieb, *Forcing the Spring: The Transformation of the Environmental Movement* (Washington, D.C.: Island Press, 1993); Ellen Maura McGurty, "From NIMBY to Civil Rights: The Origins of the Environmental Justice Movement," *Environmental History* 39 (1997): 301–323; Laura Pulido, *Environmentalism and Economic Justice: Two Chicano Struggles in the Southwest* (Tucson: University of Arizona Press, 1996); Sylvia Hood Washington, *Packing Them In: An Archaeology of Environmental Racism in Chicago, 1865–1954* (New York: Lexington, 2004).

82. http://weact.org. See also Julie Sze, "Noxious New York: The Racial Politics of Urban Health and Environmental Justice" (Ph.D. diss., New York University, 2003).

83. "Harlem Asthma Study Confirms WE ACT's Claim: If You Live Uptown, Breathe at Your Own Risk," http://weact.org/pressadvisories/2003_Apr_23.html. See also Mary E. Northridge et al., "Diesel Exhaust Exposure among Adolescents in Harlem: A Community-Driven Study," *American Journal of Public Health* 89 (1999): 998–1002.

84. Brown et al., "The Health Politics of Asthma," 461.

85. Frederica P. Perera et al., "The Challenge of Preventing Environmentally Related Disease in Young Children: Community-Based Research in New York City," *Environmental Health Perspectives* 110 (2002): 197–204; Peggy Shepard et al., "Preface: Advancing Environmental Justice through Community-Based Participatory Research," *Environmental Health Perspectives* 100 (2002): 139–140.

86. Jason Corburn, "Combining Community-Based Research and Local Knowledge to Confront Asthma and Subsistence-Fishing Hazards in Greenpoint/Williamsburg, Brooklyn, New York," *Environmental Health Perspectives* 110 (2002): 241–248.

87. Quoted in Jason Corburn, *Street Science: Community Knowledge and Environmental Health Justice* (Cambridge: MIT Press, 2005), 121.

88. Brown et al., "The Health Politics of Asthma," 461.

CHAPTER 5: ON THE HOME FRONT

1. "Charles Davies, 71; Built Air Diffusers," *New York Times,* 21 March 1969, 47. On Carrier, see Gail Cooper, *Air-Conditioning America: Engineers and the Controlled Environment, 1900–1960* (Baltimore: Johns Hopkins University Press, 1998); Marsha E. Ackermann, *Cool Comfort: America's Romance with*

Air-Conditioning (Washington, D.C.: Smithsonian Institution Press, 2002). On the history of the heating and ventilation trades, see Barry Donaldson and Bernard Nagengast, *Heat and Cold: Mastering the Great Indoors* (Atlanta: ASHRAE Publications, 1994).

2. "Isabel Beck," *New York Times,* 18 January 1990, D21.

3. Ackermann, *Cool Comfort,* 48.

4. Ruth Millard, "Hay Fever Put This Couple in Business," *American Magazine* 114 (August 1932): 69.

5. M. Murray Peshkin and Isabel Beck, "A New and Simplified Mechanical Air Filter in the Treatment of Hay Fever and Pollen Asthma," *Journal of Laboratory and Clinical Medicine* 15 (1930): 643–649; Isabel Beck, "The Treatment of Hay Fever and Pollen Asthma with Pollen Free Air," *Medical Journal and Record* (3 August 1932): 90–94.

6. *Journal of Allergy* 2 (September 1931).

7. On the history of dust as an occupational hazard, see David Rosner and Gerald Markowitz, *Deadly Dust: Silicosis and the Politics of Occupational Disease in Twentieth-Century America* (Princeton, N.J.: Princeton University Press, 1991); Christopher C. Sellers, *Hazards of the Job: From Industrial Disease to Environmental Health Science* (Chapel Hill: University of North Carolina Press, 1997). See also Joseph A. Amato, *Dust: A History of the Small and the Invisible* (Berkeley: University of California Press, 2000).

8. Milton B. Cohen, "The Prophylaxis and Treatment of Hay Fever and Asthma by Means of Pollen Filters," *Clinical Medicine and Surgery* 34 (1927): 276–278.

9. Milton B. Cohen: "The Prophylaxis and Treatment of Hay Fever and Asthma in Rooms Made Pollen and Dust Free by Means of Mechanical Filters," *Journal of Laboratory and Clinical Medicine* 13 (1927): 59–63, and "Further Observations on the Use of Filtered Air in the Diagnosis and Treatment of Allergic Conditions," *Journal of Laboratory and Clinical Medicine* 13 (1928): 963–967.

10. On the history of industrial hygiene and psychrometric chambers, see Michelle Murphy, *Sick Building Syndrome and the Problem of Uncertainty: Environmental Politics, Technoscience, and Women Workers* (Durham, N.C.: Duke University Press, 2006).

11. Simon S. Leopold and Charles S. Leopold, "Bronchial Asthma and Allied Allergenic Disorders: A Preliminary Report of a Study under Controlled Conditions of Environment, Temperature, and Humidity," *Journal of the American Medical Association* 84 (1925): 731–734.

12. Quoted in Cooper, *Air-Conditioning America,* 75.

13. These figures are taken from Cooper, *Air-Conditioning America,* and Ackermann, *Cool Comfort.*

14. William H. Welker, "Air Conditioning and Its Effects on Hay Fever and Pollen Asthma," *Heating and Ventilating* 33 (1936): 33–38.

15. See Leslie N. Gay to Mr. J. J. Nance, Manager Air Conditioning Divi-

sion, Frigidaire Corporation, 12 May 1936; Delco-Frigidaire Conditioning Corporation to Leslie N. Gay, 19 June 1936; J. J. Nance to Leslie N. Gay, 21 October 1936; LNGP.

16. Tell Nelson, B. Z. Rappaport, and William H. Welker, "The Effect of Air Filtration in Hay Fever and Pollen Asthma," *Journal of the American Medical Association* 100 (1933): 1385–1392; Tell Nelson, B. Z. Rappaport, A. G. Canar, and William H. Welker, "Air Conditioning in the Treatment of Pollen Asthma," *Heating, Piping, and Air Conditioning* 6 (1934): 329–331, 345; B. Z. Rappaport, Tell Nelson, and William H. Welker, "The Effect of Low Relative Humidity at Constant Temperature on Pollen Asthma," *Journal of Allergy* 6 (1935): 111–122; Leslie Gay, "The Treatment of Hay Fever and Pollen Asthma by Air-Conditioned Atmosphere," *Journal of the American Medical Association* (1933): 1382–1385.

17. "Halts Hay Fever by Conditioned Air," *New York Times,* 7 May 1933, 31. See also "Hay Fever: Conditioned Air Promises Relief for Victims," *Newsweek,* 13 May 1933, 26; Elmer Raymond Arn, *Relief from Hay-Fever and Other Disorders by Means of Frigidaire Air Conditioning* (Dayton, Ohio: Frigidaire Corporation, 1933).

18. Quoted in Cooper, *Air-Conditioning America,* 124.

19. J. C. Furnas, "Daniel Webster Had It Too," *Saturday Evening Post* 212 (22 July 1939): 37–39, 42.

20. T. A. Kendall and Garland Weidner, "Observations of Hay-Fever Sufferers in Air-Conditioned Room and the Relationship between Pollen Content of Outdoor Air and Weather Conditions," *Heating, Piping, and Air Conditioning* 3 (1934): 75–84.

21. "Why Do Insurance Companies Air Condition?" *Heating, Piping, and Air Conditioning* 13 (1941): 157–159.

22. Katherine Madison, "Adventures of a Sneeze Hound," *American Magazine* 131 (January 1941): 19, 112.

23. Hannah Lees, "Sensitive Souls," *Collier's* 197 (25 January 1936): 65. On Derringer, see Furnas, "Daniel Webster Had It Too."

24. John H. Ingersoll, "Ultimate Indoor Comfort," *House Beautiful* 3 (1969): 7.

25. Ibid.

26. Ibid., 7–8, 164.

27. Rachel Carson, *Silent Spring* (1962; New York: Fawcett Crest, 1964), 168.

28. Clifford Edward Clark, Jr., *The American Family Home, 1800–1960* (Chapel Hill: University of North Carolina Press, 1986), 223. See also Adam Rome, *The Bulldozer in the Countryside: Suburban Sprawl and the Rise of American Environmentalism* (Cambridge: Cambridge University Press, 2001).

29. Elaine Tyler May, *Homeward Bound: American Families in the Cold War Era* (New York: Basic Books, 1988), 165, 171.

30. Quoted in Ackermann, *Cool Comfort,* 122–123, 128–129.

31. Steven M. Spencer, "Pollen for the Sneezers," *Saturday Evening Post* 217 (26 August 1944): 105.

32. Robert A. Cooke: "Address," *New York State Journal of Medicine* 59 (15 January 1959): 290–295, and "Studies in Specific Hypersensitivity, IV: New Etiologic Factors in Bronchial Asthma," *Journal of Immunology* 7 (1922): 147–162.

33. L. Berrens, "The Allergens in House Dust," *Progress in Allergy* 14 (1970): 259–339.

34. "Asthma," *Consumer Reports,* September 1954, 433.

35. Milton B. Cohen, "Asthma and Other Allergic Conditions," *Hygeia* 7 (1929): 588.

36. "Minutes of the Meeting of the Section on Aeroallergens of the Research Council of the American Academy of Allergy, 12 December 1947," Box 160, Folder 9, p. 2, AAAAI.

37. Morris Fishbein, "Hay Fever and Sensitivity," *Saturday Evening Post* 204 (1932): 78.

38. Barbara Humphrey, "Achew! The Perils of House Dust," *Today's Health* 46 (1968): 13–15.

39. These statistics are taken from Ruth Schwartz Cowan, *More Work for Mother: The Ironies of Household Technology from the Open Hearth to the Microwave* (New York: Basic Books, 1983), 199.

40. Ellen Richards, *Euthenics, the Science of Controllable Environment: A Plea for Better Living Conditions as a First Step toward Higher Human Efficiency* (Boston: Whitcomb and Barrows, 1910).

41. On the domestic science movement, see Suellen Hoy, *Chasing Dirt: The American Pursuit of Cleanliness* (New York: Oxford University Press, 1995); Sarah A. Leavitt, *From Catharine Beecher to Martha Stewart: A Cultural History of Domestic Advice* (Chapel Hill: University of North Carolina Press, 2002); Beth Sutton-Ramspeck, *Raising the Dust: The Literary Housekeeping of Mary Ward, Sarah Grand, and Charlotte Perkins Gilman* (Athens: Ohio University Press, 2004); Nancy Tomes, *The Gospel of Germs: Men, Women, and the Microbe in American Life* (Cambridge: Harvard University Press, 1998).

42. Ellen H. Richards, *The Cost of Living as Modified by Sanitary Science,* 3rd ed. (New York: Wiley, 1910), 137–138.

43. Quoted in Tomes, *The Gospel of Germs,* 145.

44. "Allergies Alleviated," *House and Garden* 74 (1938): 67.

45. Roland Berg, "Are You Allergic?" *Scribner's Magazine* 103 (1938): 32.

46. "H. W. Hoover Dies," *New York Times,* 17 September 1954, 27; Eugene J. McGovern, "Marketing in the Vacuum Cleaner Industry: A Case Study" (MBA thesis, Brown University, 1953); http://www.hoovercompany.com/dbPages/History.asp.

47. McGovern, "Marketing in the Vacuum Cleaner Industry," 146–147.

48. See advertisement section, *Journal of Allergy* 6 (September 1935).

49. Karl D. Figley, "House Dust," *Hygeia* 24 (1946): 418–419, 450; Jerome Glaser, "The Allergic Child in Camp," *Hygeia* 23 (1945): 442–443, 464–465.

50. Arthur F. Coca, "Dust-Seal: Its Use in the Avoidance of 'House Dust' by Dust-Sensitive Persons," *Annals of Allergy* 6 (September–October 1948): 506–510, 517.

51. Arthur F. Coca, *Familial Nonreagenic Food-Allergy* (Springfield, Ill.: Charles C. Thomas, 1943). See also Sheldon G. Cohen, "Firsts in Allergy, IV: The Contributions of Arthur F. Coca, M.D. (1875–1959), *New England and Regional Allergy Proceedings* 6 (1985): 285–294; "Who's Idioblaptic?" *Time,* 1 October 1956.

52. Arthur F. Coca, "Environmental Excitants of Idioblaptic Allergy (Inhalants)," *Annals of Allergy* 6 (September–October 1948): 501–505.

53. Lawrence Galton, "K.O. for Allergy Dust," *Better Homes and Gardens* 27 (June 1949): 31, 205; "Settling Down Dust," *Newsweek* 32 (1 November 1948): 48.

54. On building the body in a safe place, see Murphy, *Sick Building Syndrome and the Problem of Uncertainty.*

55. Jerome Glaser, "The Care of the Child with Chronic Asthma," *Hygeia* 27 (1949): 174–175.

56. For an excellent discussion on the history of psychosomatic theories of asthma, see Carla C. Keirns, "Short of Breath: A Social and Intellectual History of Asthma in the United States" (Ph.D. diss., University of Pennsylvania, 2004), 169–203.

57. Irma W. Hewlett, "News about Asthma," *Parents Magazine* 22 (1947): 87.

58. Ralph Blumenthal, "Changing Hazards in the Home," *New York Times,* 30 October 1980, C1.

59. On the energy costs of suburban America, see Adam Rome, *The Bulldozer in the Countryside,* 45–86.

60. National Research Council Committee on Indoor Pollutants, *Indoor Air Pollutants* (Washington, D.C.: National Academy Press, 1981); Jane Brody, "Dangers of Indoor Air Pollution," *New York Times,* 28 January 1981, C1, C8.

61. John D. Spengler and Ken Sexton, "Indoor Air Pollution: A Public Health Perspective," *Science* 221 (1983): 9.

62. Philip Shabecoff, "Indoor Air Called Threat to Health," *New York Times,* 26 July 1981, 19.

63. Theron G. Randolph, *Human Ecology and Susceptibility to the Chemical Environment* (Springfield, Ill.: Charles C. Thomas, 1962), 10.

64. Ibid., 12–15.

65. Ibid., 121.

66. Warren T. Vaughan, *Strange Malady: The Story of Allergy* (New York: Doubleday, 1941).

67. Kimishige Ishizaka and Teruko Ishizaka, "Identification of Gamma-E

Antibodies as a Carrier of Reagenic Activity," *Journal of Immunology* 99 (1967): 1187–1198.

68. Quoted in Carl Sherman, "Chemical Allergies: Can Fresh Paint, Polyester, the Whole 20th Century Give You a Rash?" *Glamour,* November 1981, 100–101.

69. Gloria Hocham, "Allergic to Everything," *Health* 18 (1986): 53.

70. For an overview of some of these cases and disputes, see Eliot Marshall, "Immune System Theories on Trial," *Science* 234 (1986): 1490–1492; Robert Reinhold, "When Life Is Toxic," *New York Times Magazine,* 16 September 1990, 50–51, 65–70; Murphy, *Sick Building Syndrome and the Problem of Uncertainty,* 151–180.

71. "The Chemical Manufacturers Association's Environmental Illness Briefing Paper," http:www.getipm.com/articles/CMA-briefing.htm.

72. Richard B. Spohn to Raymond G. Slavin, 22 November 1983, Folder 7, Box 361, AAAAAI. See also Raymond G. Slavin to Mike Whitney, 15 July 1983; Theron Randolph to Raymond G. Slavin, 21 December 1983, Folder 7, Box 361, AAAAAI. For a sampling of scientific literature on the more public side of the dispute, see Michael H. Grieco, "Controversial Practices in Allergy," *Journal of the American Medical Association* 247 (1982): 3106–3111; Peter P. Morgan, "Should Scientists Study '20th-Century Disease,'" *Canadian Medical Association Journal* 133 (1985): 961–962; Donna Eileen Steward and Joel Raskin, "Psychiatric Assessment of Patients with '20th-Century Disease' ('Total Allergy Syndrome')," *Canadian Medical Association Journal* 133 (1985): 1001–1006; Abba I. Terr: "Environmental Illness: A Clinical Review of 50 Cases," *Archives of Internal Medicine* 146 (1985): 145–149, and "Clinical Ecology," *Journal of Allergy and Clinical Immunology,* March 1987, 423–426; Elliott Ellis, "Clinical Ecology: Myth and Reality," *Buffalo Physician,* February 1986, 17, 24–28; Doris Rapp, "Environmental Medicine: An Expanded Approach to Allergy," *Buffalo Physician,* February 1986, 16, 18–23.

73. J. A. Anderson et al., "Position Statement on Clinical Ecology," *Journal of Allergy and Clinical Immunology* 78 (1986): 270.

74. Bruce Berlow to Ann Landers, 17 October 1986, Folder 12, Box 361, AAAAAI.

75. Committee on the Assessment of Asthma and Indoor Air, *Clearing the Air: Asthma and Indoor Air Exposures* (Washington, D.C.: National Academy Press, 2000).

76. See Charles Reed to James A. Frazier, 21 July 1988, Folder 4, Box 226, AAAAAI; Andrew Pope, Roy Patterson, and Harriet Burge, eds., *Indoor Allergens: Assessing and Controlling Adverse Health Effects* (Washington, D.C.: National Academy Press, 1993). The latter study was under the auspices of the Institute of Medicine's Committe on the Health Effects of Indoor Allergens.

77. R. Voorhorst et al., "The House-Dust Mite (*Dermatophagoides pteronyssinus*) and the Allergens It Produces: Identity with the House-Dust Allergen," *Journal of Allergy* 39 (1967): 325–338; R. Voorhorst, F. Th. M. Spieksma,

and H. Varekamp, *House-Dust Atopy and the House-Dust Mite* Dermatophagoides Pteronyssinus (Trouessart 1897) (Leiden: Stafleu's Scientific Publishing, 1969); F. Th. M. Spieksma and Paul H. Dieges, "The History of the Finding of the House Dust Mite," *Journal of Allergy and Clinical Immunology* 113 (2004): 573–576.

78. J. E. M. H. van Bronswijk and R. N. Sinha, "Pyroglyphid Mites (Acari) and the House Dust Allergy: A Review," *Journal of Allergy* 47 (1971): 31–52; Thomas A. E. Platts-Mills and Martin D. Chapman, "Dust Mites: Immunology, Allergic Disease, and Environmental Control," *Journal of Allergy and Clinical Immunology* 80 (1987) 755–775.

79. Committee on the Assessment of Asthma and Indoor Air, *Clearing the Air;* Jens Korsgaard, "Changes in Indoor Climate after Tightening of Apartments," *Environment International* 9 (1983): 97–101; Ib Andersen and Jens Korsgaard, "Asthma and the Indoor Environment: Assessment of the Health Implications of High Indoor Air Humidity," *Environment International* 12 (1986): 121–127.

80. Katherine Bowman-Walzer, "I Remember Arthur F. Coca," *New England and Regional Allergy Proceedings* 6 (1985): 303–305. See also Arthur F. Coca, "Influence of Idioblastic Cigarette Sensitivity," *Annals of Allergy* 5 (1947): 458–466.

81. See Murphy, *Sick Building Syndrome and the Problem of Uncertainty,* 145–150.

82. Pope, Patterson, and Burge, eds., *Indoor Allergens.*

83. National Cancer Institute, *Health Effects of Exposure to Environmental Tobacco Smoke: The Report of the California Environmental Protection Agency,* Smoking and Tobacco Control Monograph no. 10 (Bethesda, Md.: U.S. Department of Health and Human Services, National Institutes of Health, National Cancer Institute, NIH 99-4645, 1999).

84. Committee on the Assessment of Asthma and Indoor Air, *Clearing the Air.*

85. General Accounting Office, *Indoor Pollution, Status of Federal Research Activities: Report to the Ranking Minority Member, Committee on Government Reform, House of Representatives* (Washington, D.C.: General Accounting Office, 1999).

86. William J. Fisk, "Health and Productivity Gains from Better Indoor Environments and Their Relationship with Building Efficiency," *Annual Review of Energy and Environment* 25 (2000): 537–566.

87. Linda Kulman, "Doing Battle with Dust: Choose the Right Targets and Don't Overdo It," *U.S. News and World Report* 128 (8 May 2000): 50–51.

88. "Air Cleaners," *Consumer Reports* 50 (January 1985): 7–11.

89. Nancy Sander to Albert Sheffer, 8 July 1988, "Mothers of Asthmatics," Folder 1, Box 226, AAAAI.

90. Mothers of Asthmatics, Business Plan and Budget, 1989, "Mothers of Asthmatics," Folder 2, Box 226, AAAAI.

91. http://partners.guidestar.org.

92. Michael D. Cabana et al., "Parental Management of Asthma Triggers within a Child's Environment," *Journal of Allergy and Clinical Immunology* 114 (2004): 352–357.

CHAPTER 6: AN INHALER IN EVERY POCKET

1. Greg Critser, *Generation Rx: How Prescription Drugs Are Altering American Lives, Minds, and Bodies* (Boston: Houghton Mifflin, 2005), 116.

2. http://advair.com. Sales figures for Advair are from GlaxoSmithKline, *Annual Report,* 2005 and 2001.

3. http://gw-flovent.com; GlaxoSmithKline, *Annual Report,* 2005.

4. This figure is based on Schering-Plough's annual reports from 1993 to 2002. For an insightful look into the making of Claritin into a blockbuster drug, see Stephen S. Hall, "Prescription for Profit," *New York Times Sunday Magazine,* 11 May 2001, 40–45, 59, 91–92, 100.

5. On one approach to an environmental history of the body, see Christopher Sellers, "Thoreau's Body: Towards an Embodied Environmental History," *Environmental History* 4 (1999): 486–514.

6. These figures are from *Historical Tables: Budget of the United States Government, Fiscal Year 2005* (Washington, D.C.: U.S. Government Printing Office, 2004), 45–46.

7. My thanks to Paul Erickson for the material contained in this and the following paragraph. Paul's knowledge of the history of postwar economic planning in the United States is unsurpassed. See also Michael Sherry: *Preparing for the Next War: American Plans for Postwar Defense, 1941–1945* (New Haven: Yale University Press, 1977), and *In the Shadow of War: The United States since the 1930's* (New Haven: Yale University Press, 1995); Aaron Friedberg, *In the Shadow of the Garrison State: America's Anti-Statism and Its Cold War Grand Strategy* (Princeton, N.J.: Princeton University Press, 2000).

8. Leo Nejelski, "A Diagnosis for the Drug Industry," *American Druggist,* May 1947, 74.

9. Lizabeth Cohen, *A Consumers' Republic: The Politics of Mass Consumption in Postwar America* (New York: Knopf, 2003). On the history of the drug industry and consumer society in twentieth-century America, see Jackson Lears, *Fables of Abundance: A Cultural History of Advertising in America* (New York: Basic Books, 1994); Nancy Tomes, "The Great American Medicine Show Revisited," *Bulletin of the History of Medicine* 79 (2005): 627–663.

10. Sales figures are from Nathan Wishnefsky, *Antihistamines Industry and Product Survey* (New York: Chemonomics, 1950). In August 1952, antihistamines accounted for 10.3 percent of all prescriptions filled, making them third most popular prescription drug for the first time. See "Prescription Trends," *American Druggist,* 4 August 1952, 27–28. Antibiotic sales figures are from "Now What?" *American Druggist,* October 1949, 72.

11. "The Cold War in Antihistaminics," *Chemical and Engineering News* 28 (1950): 846–848, 858.

12. H. H. Dale and P. P. Laidlaw, "The Physiological Action of ß-iminazolylethylamine," *Journal of Physiology* 41 (1911): 318; Henry Dale, "The Pharmacology of Histamine: With a Brief Survey of Evidence for Its Occurrence, Liberation, and Participation in Natural Reactions," *Annals of the New York Academy of Sciences* 50 (1950): 1017–1028. See also E. M. Tansley, "Henry Dale, Histamine and Anaphylaxis: Reflections on the Role of Chance in the History of Allergy," *Studies in History and Philosophy of Biological and Biomedical Sciences* 34 (2003): 455–472.

13. S. M. Feinberg, S. Malkiel, and A. R. Feinberg, *The Antihistamines* (Chicago: Year Book Publishers, 1950); L. W. Giellerup, "The Culprit Is Histamine," *Scientific American* 169 (August 1943): 68–69.

14. Samuel M. Feinberg, with the collaboration of Oren C. Durham and Carl A. Dragstedt, *Allergy in Practice,* 2nd ed. (Chicago: Year Book Publishers, 1946), 801–815.

15. The Benadryl story is well told in Steven M. Spencer, "New Hope for the Allergic," *Saturday Evening Post* 218 (20 April 1946): 21, 121–122. See also Leo H. Criep, "Conquering Your Allergies with Drugs," *Hygeia* 25 (1947): 98–99, 158.

16. "Announces Hay Fever Drug," *New York Times,* 5 April 1946, 36; William Laurence, "New Chemical Relieves Hay Fever and Hives, Scientists Are Told," *New York Times,* 27 October 1946, 1. See also Albert Deutsch, "New 'Cures' for Hay Fever," *Science Digest* 20 (1946): 10–15. This number is based on the comprehensive list in Feinberg, Malkiel, and Feinberg, *The Antihistamines,* 220–231.

17. "Pyribenzamine. . . ," *Journal of Allergy* 118 (1947): 3; "Prescribed the 'Year Round . . . ," *American Druggist,* September 1947, 115.

18. "An 'Ill Wind' That Blows Good Business In . . . ," *American Druggist,* August 1947, 95.

19. For a brief company history of G. D. Searle, see Tom Mahoney, *The Merchants of Life: An Account of the American Pharmaceutical Industry* (New York: Harper, 1954), 144–152.

20. See Austin Smith to Leslie N. Gay, 21 July 1949, "Personal, ACTH" Folder, Box 2, LNGP.

21. Leslie N. Gay to Mark Nickerson, 5 May 1950, "V-A Science" Folder, Box 2, LNGP.

22. The story was widely retold in Gay's private correspondence and in the popular press. See, e.g., Horace Sutton, "So You Were Going to Get Seasick," *Saturday Review of Literature* 32 (7 May 1949): 41–42.

23. Leslie N. Gay, Paul E. Carliner, and Joseph E. Moore, "The Prevention and Treatment of Motion Sickness," *Transactions of the Association of American Physicians* 62 (1949): 202.

24. Stock prices are from Leslie Gay to Alan Chesney, 9 April 1953, "Gay,

Leslie N. Bio" Folder, LNGP. On the unanticipated publicity, see "Behind the Scenes with Dramamine, *The Searle Circle* 12 (n.d.): 1–5, in "II-B Searle" Folder, Box 1, LNGP.

25. A. C. Bratton to Leslie Gay, 22 March 1949, "II-A Searle 1949" Folder, Box 1, LNGP.

26. Irwin Winter to Leslie Gay, 14 April 1949, "II-A Searle 1949" Folder, Box 1, LNGP.

27. Leslie N. Gay, "The Development of Dramamine," *Postgraduate Medicine,* January 1956, 80–85; C. C. Shaw, "Dramamine Trials in the United States Navy," *Military Surgeon* 106 (1950): 441–449.

28. Irwin Winter to Leslie Gay, 29 November 1949, "II-A Searle 1949" Folder, Box 1, LNGP.

29. Based on Mahoney, *The Merchants of Life,* 145, and "Prospectus, G. D. Searle and Co.," 1950, "II-B Searle, Searle Circle" Folder, Box 1, LNGP.

30. Howard A. Rusk, "Holiday Fun Insurance," *New York Times,* 1 July 1956, 33.

31. "A Pleasant Relief from Hay Fever," *Journal of Allergy* 19 (1948): 13.

32. N. D. Fabricant, "A Plan to Eradicate the Common Cold," *Eye, Ear, Nose and Throat Monthly* 25 (1946): 241–243.

33. Figure quoted in Wishnefsky, *Antihistamines Industry and Product Survey,* 71.

34. Fabricant, "A Plan to Eradicate the Common Cold," 241.

35. J. M. Brewster, "Benadryl as a Therapeutic Agent in the Treatment of the Common Cold," *U.S. Naval Medical Bulletin* 47 (1947): 810–811.

36. J. M. Brewster, "Antihistamine Drugs in the Therapy of the Common Cold," *U.S. Naval Medical Bulletin* 49 (1949): 1–11.

37. M. M. Kessler, "Hydrillin for the Common Cold," *Journal of the Medical Society of New Jersey* 47 (1950): 29–30.

38. "Allergy in Epidemiology of Common Cold," *Journal of the American Medical Association* 141 (1949): 138. The editorial was based in large part on the findings reported in N. Fox and G. Livingston, "Role of Allergy in the Epidemiology of the Common Cold," *Archives of Otolaryngology* 49 (1959): 575–586.

39. Brewster, "Antihistaminic Drugs in the Therapy of the Common Cold," 8.

40. Paul De Kruif, "Is This, at Last, Goodbye to the Common Cold?" *Reader's Digest* 55 (December 1949): 16–18.

41. "Cold War Gets Warmer," *Business Week,* 17 December 1949, 68–69; "The Cold Rush," *Modern Packaging,* January 1950, 78–80; "The Cold War in Antihistaminics," *Chemical and Engineering News* 28 (13 March 1950): 846–848, 858. On the Cold Control Plan, see photo in *American Druggist,* March 1950, 50.

42. See "MDs Hit Unlimited Anthistamine Use," *Drug Trade News,* 9 January 1950, 35.

43. "Status Report on Antihistaminic Agents in the Prophylaxis and Treatment of the Common Cold," *Journal of the American Medical Association* 142 (1950): 569.

44. Ibid., 567.

45. Quoted in Wishnefsky, *Antihistamines Industry and Product Survey,* 103.

46. Herman Bundesen, "Dangers for Youngsters in Antihistamines," *Ladies' Home Journal,* June 1950, 192, 194.

47. "Antihistamine Ad Claim Suits Ended by FTC," *Drug Trade News* 25 (26 June 1950): 1, 29.

48. "Let's Win This Cold War," *Collier's* 125 (22 April 1950): 90.

49. "Marketing of Cold Item Links O-T-C, Proprietary Techniques," *American Druggist,* 26 August 1957, 16.

50. A fortnightly survey conducted by the *American Druggest* in May 1956 found that specialists wrote 43 percent of prescriptions for antiallergic drugs. The national average for all drugs prescribed by specialists was 23.5 percent. See "Specialists Prescribe 42.9% of Anti-Allergic Rxs; 31-Month High," *American Druggist,* 4 June 1956, 28.

51. These statistics were compiled from R. A. Gosselin, *Therapeutic Category Report, Ten-Year Trend, 1959–1968* (Dedham, Mass.), available from the AIHP.

52. Wishnefsky, *Antihistamines Industry and Product Survey,* 116.

53. On the history of bronchodilators and their use in asthma treatment, see Eric K. Chu and Jeffrey M. Drazen, "Asthma: One Hundred Years of Treatment and Onward," *American Journal of Respiratory and Clinical Care Medicine* 171 (2005): 1202–1208; James P. Kemp, "Adrenergic Bronchodilators, Old and New," *Journal of Asthma* 20 (1983): 445–453. See also Parke-Davis, *Adrephine,* Kremer Files C35(a) I, "Parke-Davis and Co. 1930–1940" Folder, AIHP.

54. For a brief overview of the development of aerosol technology, see "39 Aerosol-Packed Drug Items Are Now on Market; Many More on the Way," *American Druggist,* 13 August 1956, 30, 33–34.

55. "99% of Druggists Sell Aerosol Lather," *American Druggist,* 31 January 1955, 66.

56. "Though Little-Known, Aerosol Drug Items Rang up $3,500,000 in Sales Last Year," *American Druggist,* 7 October 1957, 86; Paul Kallós and Liselotte Kallós-Deffner, "Medihaler®: A New Device for Use in the Symptomatic Treatment of Bronchial Asthma," *International Archives of Allergy* 15 (1959): 343–349.

57. Sales figures were compiled from *National Prescription Audit: Therapeutic Category Report* for 1959–1972. Courtesy of AIHP.

58. "Increasing Deaths from Asthma," *British Medical Journal* 1 (1968): 329–330; F. E. Speizer, R. Doll, and P. Heaf, "Observations on Recent Increases in Mortality from Asthma," *British Medical Journal* 1 (1968): 335–339;

F. E. Speizer et al., "Investigation into Use of Drugs Preceding Death from Asthma," *British Medical Journal* 1 (1968): 339–343.

59. For an excellent history of this debate, see Mark Jackson, *Allergy: The History of a Modern Malady* (London: Reaktion, 2006), 103–116. My thanks to Mark for sharing this chapter with me in its unpublished form.

60. W. H. W. Inman and A. M. Adelstein, "Rise and Fall of Asthma Mortality in England and Wales in Relation to Use of Pressurized Aerosols," *The Lancet*, 9 August 1969, 279–284; "Deaths from Asthma in Young People," *British Medical Journal* 1 (1972): 459.

61. Quoted in Milton Silverman and Philip R. Lee, *Pills, Profits, and Politics* (Berkeley: University of California Press, 1974), 249.

62. Paul D. Stolley, "Asthma Mortality: Why the United States Was Spared an Epidemic of Deaths Due to Asthma," *American Review of Respiratory Disease* 105 (1972): 883–890; "Asthma Deaths: A Question Answered," *British Medical Journal* 4 (1972): 443–444.

63. M. Coleman Harris, "Are Bronchodilator Aerosol Inhalations Responsible for an Increase in Asthma Mortality?" *Annals of Allergy* 29 (1971): 250–256; Irvin Caplan and John T. Haynes, "Complications of Aerosol Therapy in Asthma," *Annals of Allergy* 27 (1969): 65–69; Thomas E. Van Metre, Jr., "Adverse Effects of Inhalation of Excessive Amounts of Nebulized Isoproterenol in Status Asthmaticus," *Journal of Allergy* 43 (1969): 101–113.

64. Based on *National Prescription Audit*, AIHP. See also Harris, "Are Brochodilator Aerosol Inhalations Responsible for an Increase in Asthma Mortality?"

65. Murray Dworetzky, "Changing Concepts of the Asthma Problem," *Journal of Allergy* 43 (1969): 320–321. See also C. J. Falliers, "Pharmacology and Logic," *Journal of Allergy* 44 (1969): 59.

66. Kimishige Ishizaka and Teruko Ishizaka, "Identification of Gamma-E Antibodies as a Carrier of Reagenic Activity," *Journal of Immunology* 99 (1967): 1187–1198; Sami L. Bahna, "A 21-Year Solute to IgE," *Annals of Allergy* 62 (1989): 471–478; Teruko Ishizaka, "IgE and Mechanisms of IgE-Mediated Hypersensitivity," *Annals of Allergy* 48 (1982): 313–319.

67. For semipopular accounts of how the immune landscape was seen by the 1980s, see Paul D. Bussieret, "Allergy," *Scientific American* 247 (1982): 86–95; Susan V. Lawrence, "Recent Advances in Hay Fever Research Are Nothing to Sneeze At," *Smithsonian* 15 (1984): 100–110; Lawrence M. Licthenstein, "Allergy and the Immune System," *Scientific American* 269 (1993): 85–93. On changing concepts of the immune system in American culture, see Emily Martin, *Flexible Bodies: The Role of Immunity in American Culture from the Days of Polio to the Age of AIDS* (Boston: Beacon Press, 1995).

68. L. A. Bosco, D. E. Knapp, B. Gerstman, and C. F. Graham, "Asthma Drug Therapy Trends in the United States, 1972 to 1985," *Journal of Allergy and Clinical Immunology* 80 (1987): 398–402.

69. See Schering-Plough, *Annual Reports*, 1982–1989.

70. For an excellent account of how changing asthma treatments are tied to changing definitions of asthma as a disease, see Carla C. Keirns, "Short of Breath: A Social and Intellectual History of Asthma in the United States" (Ph.D. diss., University of Pennsylvania, 2004), 211–215.

71. *Running Hard, Breathing Easy: The Jeanette Bolden Story* (Schering Corporation, 1982). Available through the National Library of Medicine. The film was distributed with the help of the Allergy and Asthma Foundation, a volunteer patient advocacy/education group with close ties to the American Academy of Allergy and Immunology. See Asthma and Allergy Foundation of America, *1982 Report to the American People.*

72. L. Podolsky, "Pound of Prevention, Breathe Easier," *World Tennis* 31 (1984): 62; Pearson quoted in J. Silberner, "Olympic Asthmatics Breathing Easy," *Science News* 128 (1985): 278;

73. Horace Baldwin to Perry Stucker, 5 March 1951, Folder 47, Box 446, AAAAAI.

74. Edwin Hays to Carl Abersman, 2 January 1957, Folder 13, Box 122, AAAAAI.

75. "Joint Marketing Planning Committee Meeting," Folder 28, Box 449, AAAAAI.

76. Alan Leahigh to Task Force on Marketing Allergists, 3 June 1986, Folder 28, Box 449, AAAAAI; Marybeth Brennan to Task Force on Marketing Allergists, Folder 28, Box 449, AAAAAI.

77. Quoted in Bill Stokes, "Pollen Puts a Pall on Kiss of Spring," *Chicago Tribune,* 15 May 1986, sec. 2, p. 3.

78. Marcia Angell, *The Truth about the Drug Companies: How They Deceive Us and What to Do about It* (New York: Random House, 2004).

79. "Alupent MDI Safety Overview," Folder 8, Box 352, AAAAAI.

80. FDA, Docket No. 94N-0232, p. 3.

81. Leslie Hendeles, "Nonprescription Sale of Inhaled Metaproterenol—Déjà vu," *New England Journal of Medicine* 310 (1984): 207–208.

82. Quoted in P. M. Boffey, "Asthma Drug Move Called Hazardous—Critics Warn FDA Approval of Sale of Inhaler without Prescription Is Mistake," *New York Times,* 14 April 1983, Y7.

83. Michael Sly et al., "Adverse Effects and Complications of Treatment with Beta-Adrenergic Agonist Drugs," *Journal of Allergy and Clinical Immunology* 75 (1985): 443–449.

84. "American Academy of Allergy and Immunology, Ad Hoc Committee on Metaproterenol OTC, July 8, 1985," p. 5, Folder 9, 2/2, Box 352, AAAAAI.

85. Paul Hanaway to Theodore Gotthelf, 15 July 1985, Folder 8, Box 352, AAAAAI.

86. Quoted in "Prescription Should Be Required for Beta Agonist Drugs," *Family Practice News,* 15–31 July 1985, 20.

87. Thomas Tiedt to Albert Sheffer, 28 August 1985, Folder 9, Box 352, AAAAAI.

88. "Metaproterenol MDI OTC Sub-Committee Report," Folder 8, Box 352, AAAAAI. On the change in the final report, see I. Leonard Bernstein to John Salvaggio, 5 November 1985, Folder 8, Box 352; Thomas Tiedt to Paul Hanaway, 31 October 1985, Folder 8, Box 114, AAAAAI.

89. John Anderson to John Salvaggio, 7 November 1985, Folder 8, Box 114, AAAAAI.

90. Donald McNeil to Executive Committee Members, 19 November 1985, Folder 9, Box 352, AAAAAI.

91. Thomas Tiedt to Albert Sheffer, 10 January 1986, Folder 1, Box 117, AAAAAI. See also Albert Sheffer to Theodore Gotthelf, 31 December 1985, Folder 13, Box 208, AAAAAI.

92. Peter Wise to FDA, 18 April 1986, Folder 9, 2/2, Box 352, AAAAAI. See also Gil Mott to Albert Sheffer, 30 April 1986, Folder 9, 2/2, Box 352, AAAAAI.

93. Albert Sheffer to Irving A. Tabachnick, 31 December 1985, Folder 6, Box 236, AAAAAI.

94. Albert Sheffer to Thomas Tiedt, 17 January 1986, Folder 13, Box 208, AAAAAI.

95. Centers for Disease Control, "Asthma—United States, 1980–1987," *Morbidity and Mortality Weekly Report* 39 (1990): 493–497; P. J. Gergen, D. I. Mullaly, and R. Evans III, "National Survey of Prevalence of Asthma among Children in the United States, 1976 to 1980," *Pediatrics* 81 (1988): 1–7.

96. National Asthma Education Program, *Executive Summary: Guidelines for the Diagnosis and Management of Asthma* (Bethesda, Md.: U.S. Department of Health and Human Services, 1991); NIH Publication no. (PHS) 91-3042.

97. See, e.g., Marjorie Roberts, "Attack on Asthma," *U.S. News and World Report* 110 (4 March 1991): 60–61; Bill LeGro with Camela London, "Take Control of Your Asthma Now!" *Prevention* 44 (1992): 60–65.

98. On the changing definition of asthma, see Robert A. Barbee and Shirley Murphy, "The Natural History of Asthma," *Journal of Clinical Allergy and Immunology* 102 (1998): S65–S72.

99. "Draft Excerpt from March 12, 1988, Meeting of AAAI Executive Committee," Folder 3, Box 123, AAAAAI. See also Philip Satow to James Kemp, 13 March 1987, Folder 1, Box 123; Howard Solomon to James Kemp, 2 April 1987, Folder 1, Box 123; Albert Sheffer to Elliott Ellis, 3 May 1987, Folder 1, Box 123; R. Michael Sly to Albert Sheffer, 25 May 1987, Folder 1, Box 123; Philip Satow to James Kemp, 18 June 1987, Folder 1, Box 123; James Kemp to Members of the Committee on Drugs, 9 October 1987, Folder 1, Box 123; James Audibert to James Kemp, 11 February 1988, Folder 3, Box 123; Howard Solomon to Albert Sheffer, Folder 3, Box 123; Miles Weinberger to Executive Committee, Folder 3, Box 123; all in AAAAAI.

100. Rick Iber to Brian Smith, 28 December 1988, Folder 15, Box 218, AAAAAI.

101. Forest Laboratories, *Annual Report 1991,* 6.

102. Timothy Ferris et al., "Are Minority Children the Last to Benefit from a New Technology? Technology Diffusion and Inhaled Corticosteroids for Asthma," *Medical Care* 44 (2006): 81–86.

EPILOGUE

1. René Dubos, *The Mirage of Health: Utopias, Progress, and Biological Change* (New York: Harper, 1959), 136, 128.

2. René Dubos, "The Spaceship Earth," *Journal of Allergy* 44 (1969): 1–9.

3. F. M. Burnet, *Biological Aspects of Infectious Disease* (Cambridge: Cambridge University Press, 1940), 4. On the history of disease ecology, see Warwick Anderson, "Natural Histories of Infectious Disease: Ecological Vision in Twentieth-Century Biomedical Science," *Osiris* 2nd ser. 19 (2004): 39–61; Gregg Mitman, "In Search of Health: Landscape and Disease in American Environmental History," *Environmental History* 10 (2005): 184–209.

4. Dubos, "The Spaceship Earth," 2.

5. Richard Levins and Cynthia Lopez, "Toward an Ecosocial View of Health," *International Journal of Health Services* 29 (1999): 261–293.

6. Dubos, *The Mirage of Health,* 99.

INDEX

Page numbers in italics refer to figures.